Santiago Borja

PICTURES BY

#THE STORMPILOT

Santiago Borja

PICTURES BY

#THE STORMPILOT

TEXT BY
Michaela Koschak

teNeues

#Hard to believe all you see here is nothing but light and water

Contents

Preface
Santiago Borja

The most intensive feeling comes right at the start. When the aeroplane suddenly accelerates and I'm pressed into my seat in the cockpit. First the front, and then the back wheels lose contact with the ground. No matter how often I experience this moment, it never feels routine.

My name is Santiago Borja. I'm a pilot for one of the world's major airlines. I've been flying long haul across the planet for years now and for many of those years, I've been taking photographs of exiting things I see from the cockpit: towering piles of clouds, storms and lightning, breathtaking colours in the morning and the evening, spectacularly starry skies and the lights of cities snaking out across the dark earth at night. The view from high up is truly unique: it opens up new, almost otherworldly perspectives that most of us only rarely get to see.

I would like you to join me in viewing the world from these perspectives. We're going to get right up close to storms floating through the sky at 40,000 feet and see fantastic sheet lighting inside banks of clouds. The photos are accompanied by gripping descriptions and explanations of the meteorological phenomena I get to see from my cockpit. Find out how clouds or lightning actually happen, what a storm really is, and what happens when lightning strikes; learn about the tremendous force of hurricanes and about where the earth's

Vorwort
Santiago Borja

Das intensivste Gefühl ist der Start. Wenn das Flugzeug plötzlich beschleunigt, es mich in meinem Sitz im Cockpit nach hinten drückt, wenn zuerst Vorder- und dann Hinterreifen die Bodenhaftung verlieren – dieses Erlebnis wird nie zur Routine.

Mein Name ist Santiago Borja, ich bin Pilot und arbeite bei einer großen, global vernetzten Fluggesellschaft. Seit vielen Jahren fliege ich rund um den Globus, seit vielen Jahren fotografiere ich dabei aus meinem Cockpit die spannendsten Motive: Wolkentürme, Gewitterstürme, Blitze, atemberaubendes Morgen- oder Abendrot, spektakuläre Sternenhimmel und die verzweigten Stadtlichter am Erdboden. Der Blick von oben ist einzigartig. Er eröffnet uns neue, beinah außerirdische Welten, die den meisten von uns fast ihr ganzes Leben lang verborgen bleiben.

In diese Welt möchte ich Sie mit diesem Fotoband mitnehmen. In die nächste Nähe von Gewitterwolken 12.000 Meter über dem Erdboden, zu fantastischen Wetterleuchten im Inneren von Wolkenbänken. Spannende Texte zu den Wetterphänomenen, die ich aus meinem Cockpit heraus beobachten kann, machen dieses Buch noch kurzweiliger: Wie entstehen eigentlich Wolken oder Blitze, was ist ein Gewitter und was geschieht bei einem Blitzeinschlag? Wie gewaltig können Hurrikans werden und wo beginnt eigentlich der Weltraum?

Préface
Santiago Borja

C'est au décollage que je ressens les sensations les plus fortes, quand l'accélération soudaine de l'avion me plaque contre le dossier de mon siège dans le cockpit, quand les roues avant puis arrière perdent le contact avec le sol. Cette expérience ne cède jamais à la routine.

Je m'appelle Santiago Borja et je suis pilote de d'une grande compagnie aérienne opérant sur un réseau international. Depuis de nombreuses années, je fais le tour du monde par les airs. Depuis tout aussi longtemps, je photographie de mon cockpit les motifs les plus captivants qui soient : tours de nuages, tempêtes orageuses, éclairs, aubes et crépuscules à vous couper le souffle, voûtes étoilées spectaculaires et motifs ramifiés formés par les lumières des villes au sol. À nulle autre pareille, la vue aérienne dévoile des mondes nouveaux, presque extraterrestres, que la plupart d'entre nous n'ont guère l'occasion de voir dans toute leur vie.

C'est dans ce monde que je voudrais vous emmener par le biais de cet ouvrage de photographie, tout près des nuages orageux, à 12 000 mètres au-dessus du sol, pour admirer de spectaculaires éclairs de chaleur au cœur de bancs de nuages. Vous apprécierez d'autant plus les images qu'elles sont accompagnées de textes captivants sur les phénomènes météorologiques que j'observe de mon

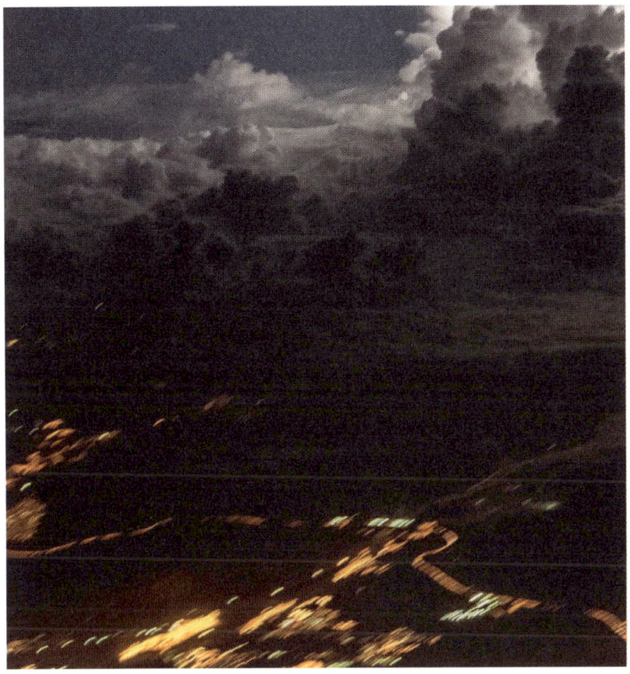

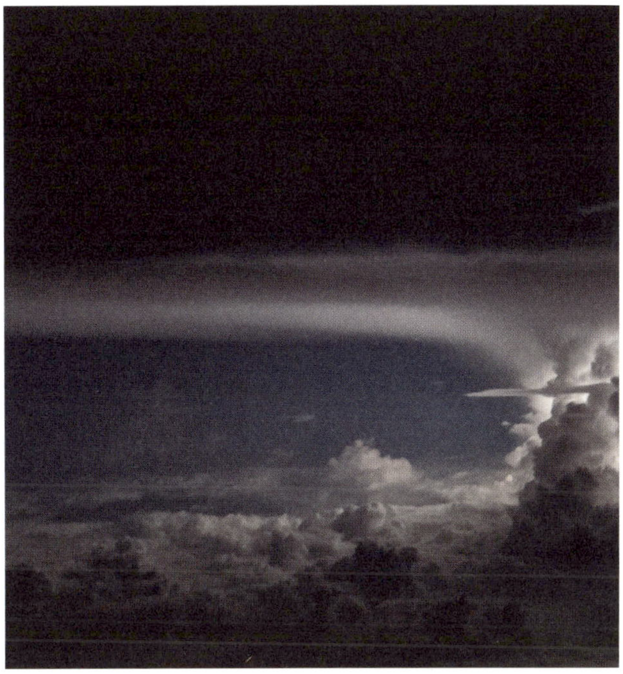

atmosphere stops and space begins. Written by Michaela Koschak, a weather forecaster on German television with a degree in meteorology, these texts have the authority that comes with genuine expertise.

You can now often find my photos as premium prints in galleries and living rooms across the world; I've been fortunate enough to win prizes in many countries. This book is a collection of some of my best works. Reading it means joining me at my desk above the clouds and taking a look at the world from my perspective – from a perspective in which the distinction between "up" and "down" can disappear at any point.

Santiago Borja
#the stormpilot

Geschrieben hat diese Texte Michaela Koschak, Journalistin beim deutschen Fernsehen und Diplom-Meteorologin – eine Expertin auf ihrem Gebiet.

Meine Bilder haben mittlerweile viele internationale Preise gewonnen und hängen als hochwertige Fotoprints in Galerien und Wohnzimmern weltweit. Dieser Bildband sammelt nun meine besten Arbeiten. Lassen Sie sich mitnehmen an meinen Arbeitsplatz über den Wolken – in eine Welt, wo der Unterschied zwischen oben und unten jeden Moment zu verschwimmen scheint.

Santiago Borja
#the stormpilot

cockpit. Comment se forment au juste les nuages ou les éclairs ? Qu'est-ce qu'un orage ? Et que se passe-t-il lors d'un impact de foudre ? Quels sont les possibles dangers des ouragans ? Où l'espace commence-t-il ? Pour répondre à ces questions, les textes ont été rédigés par une spécialiste en la matière, à savoir Michaela Koschak, journaliste à la télévision allemande et météorologue diplômée.

Mes photographies ont été récompensées par de nombreuses distinctions internationales. Des tirages de qualité sont exposés dans des galeries, ainsi que chez des particuliers, dans le monde entier. Cet ouvrage réunit mes meilleurs travaux. Venez vous joindre à moi dans mon poste de travail au-dessus des nuages, dans un univers où la frontière entre haut et bas semble s'estomper à tout moment.

Santiago Borja
#the stormpilot

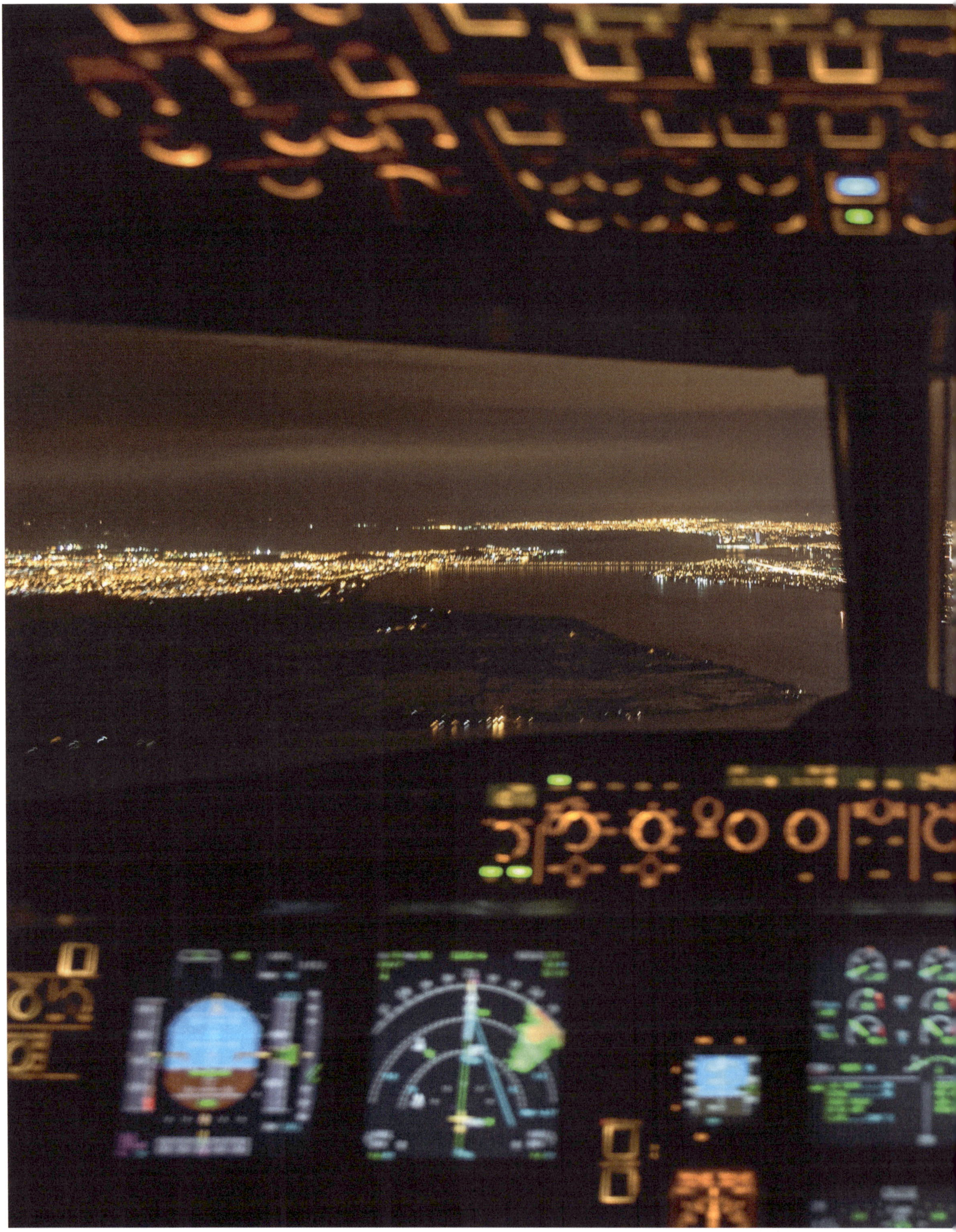

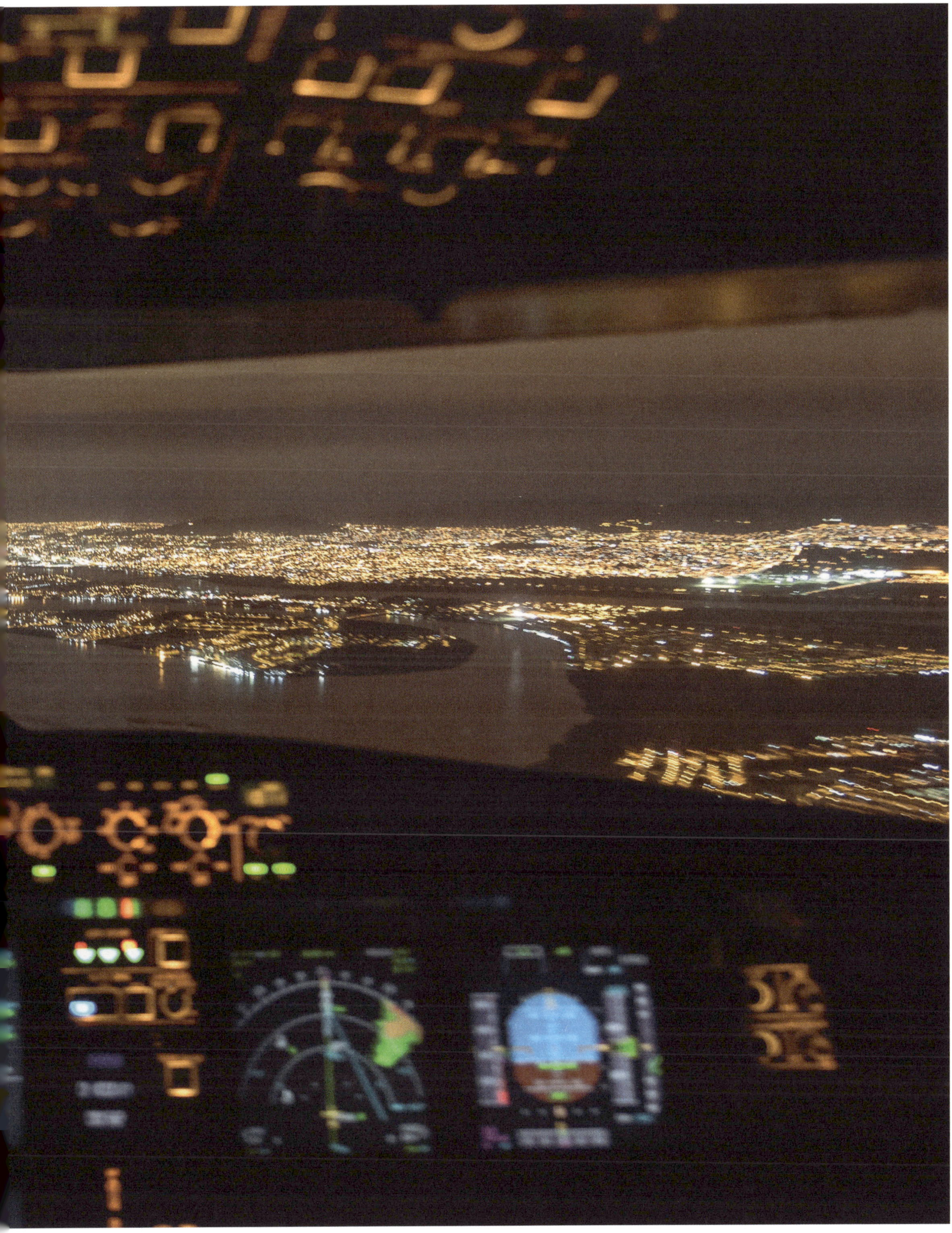

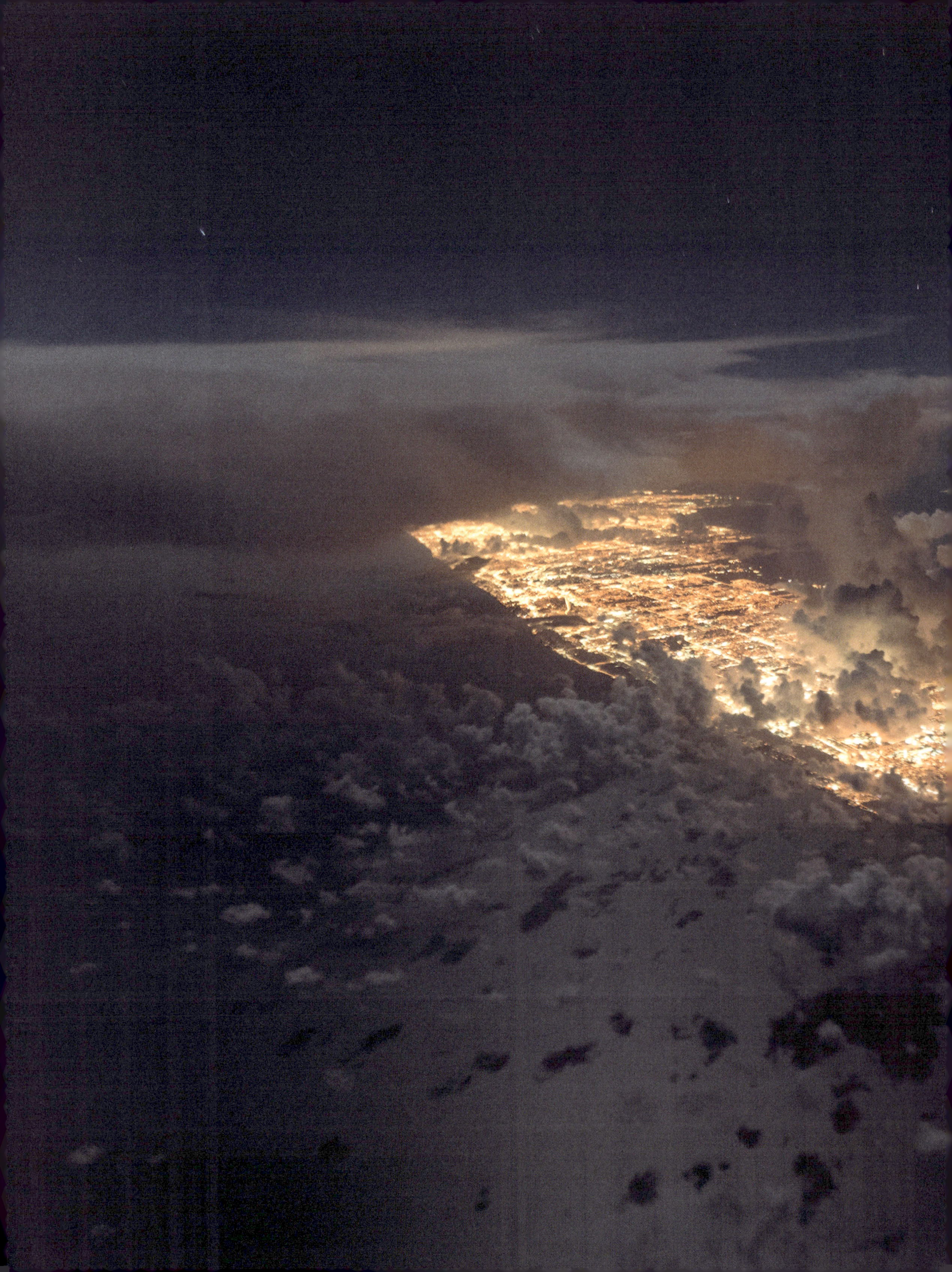

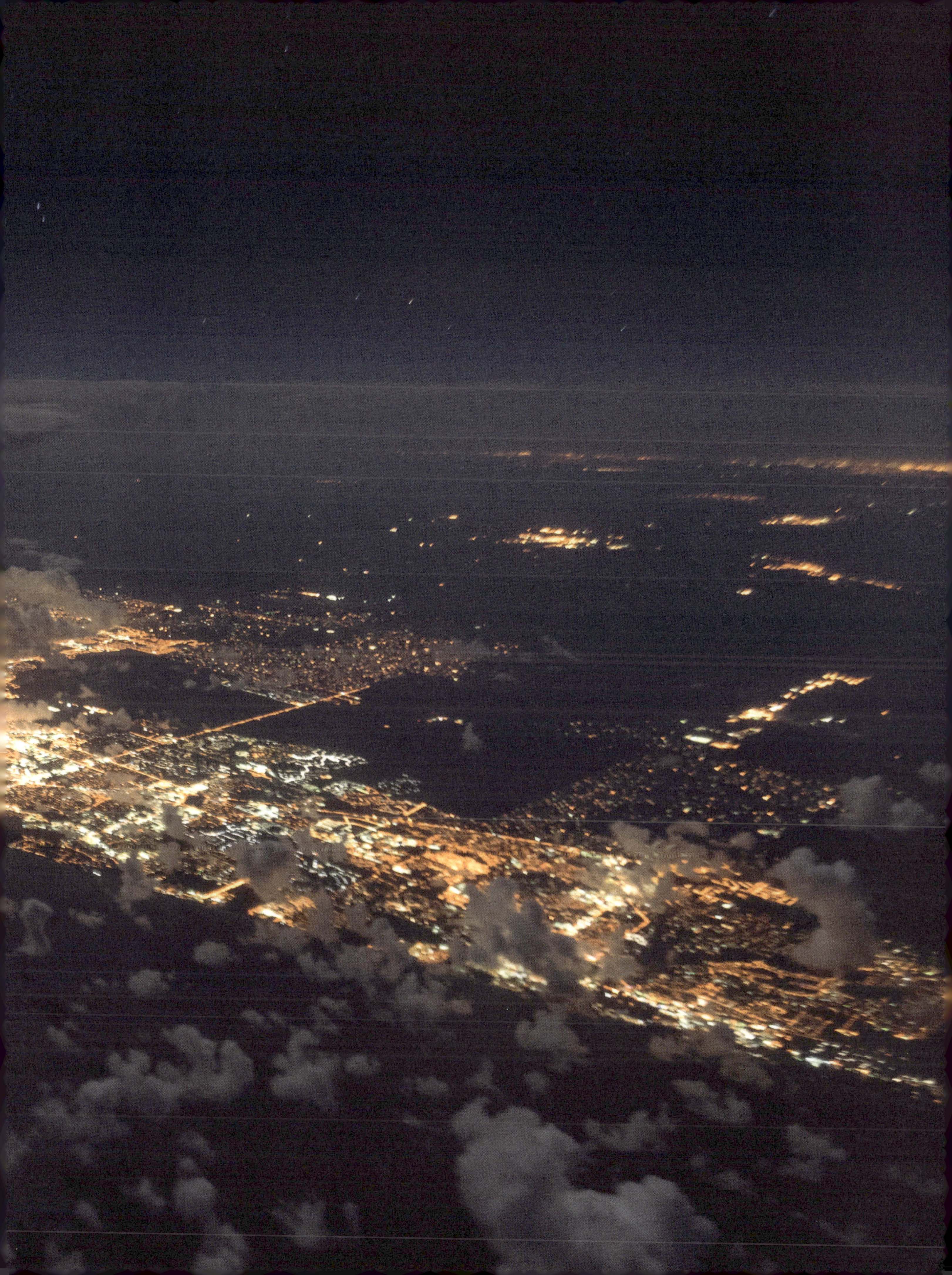

#These views never get old

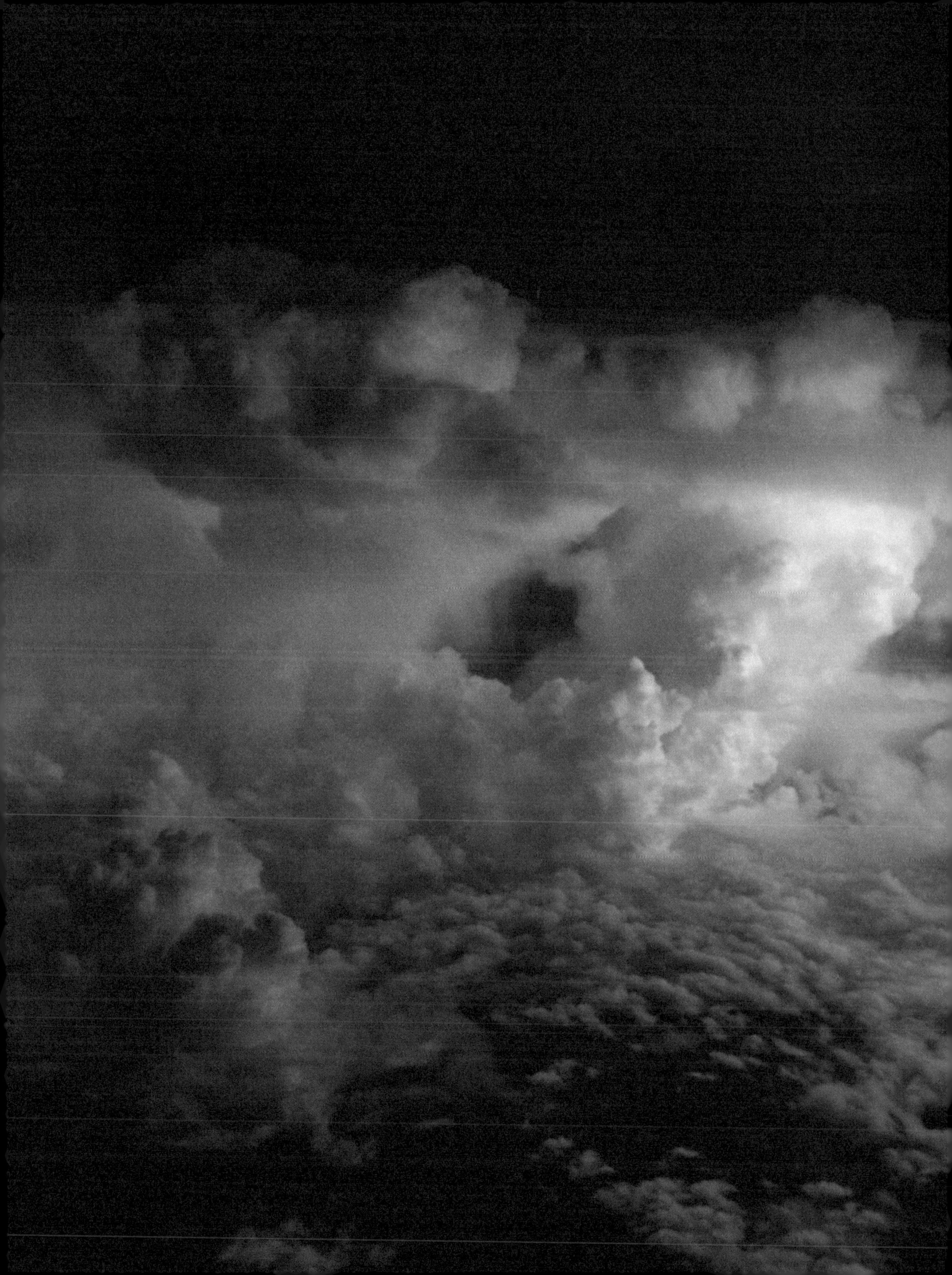

What is a cloud?

In simple terms, a cloud can best be compared with a sponge in the bathtub. When dry, it weighs almost nothing; if it absorbs a lot of water, though, it becomes heavier and heavier until, fully saturated, it starts to drip. Although clouds are of course not sponges, but rather collections of very finely grained droplets of water or ice crystals in the atmosphere, they work in a similar way: when a cloud gets too heavy, it starts to produce rain or, if the conditions are given, snow.

HOW DO CLOUDS ACTUALLY FORM IN THE FIRST PLACE, THOUGH?

How do clouds actually form in the first place, though? The main driver behind cloud formations is (as with so many other meteorological phenomena), the sun. Its rays cause water to evaporate from the sea, from lakes and rivers and from every pool or puddle of water. The water vapour rises skywards with the currents of warm air and the higher the temperature, the more water molecules can make their way upwards. At higher altitudes, however, the air temperature drops; as cold air cannot hold as many water molecules as warm air, they start to run out of room. At the point which meteorologists talk about as "saturation", the water condenses

Was ist eine Wolke?

Am ehesten kann man eine Wolke mit einem Schwamm in der Badewanne vergleichen. Wenn der Schwamm trocken ist, hat er kaum Gewicht. Saugt er sich aber voll mit Wasser, wird er schwerer und schwerer. Hat er sich schließlich komplett vollgesaugt, dann fängt er an zu tropfen. Wolken funktionieren ganz ähnlich. Sie sind Ansammlungen sehr feiner, winzig kleiner Wassertröpfchen oder Eiskristalle in der Atmosphäre. Wenn die Wolke allerdings zu schwer wird, dann beginnt es zu regnen oder – je nach Wetterlage – zu schneien.

ABER WIE ENTSTEHEN WOLKEN EIGENTLICH?

Hauptverantwortlicher ist – wie fast immer beim Wetter – die Sonne. Wenn sie scheint, verdunstet aus unseren Meeren, Seen, Flüssen und Pfützen das Wasser. Die Wasserteilchen steigen mit der warmen Luft in den Himmel – sie verdunsten. Je wärmer dabei die Luft, desto mehr Wasserteilchen können nach oben steigen. In großer Höhe aber wird die Luft natürlich wieder kälter, und kalte Luft kann weniger Wasserteilchen aufnehmen als warme. Es wird also immer enger für die Teilchen, der Platz immer spärlicher. Bis die Luft – wie wir Meteorologen sagen – „gesättigt" ist. Das ist dann genau der Moment, in dem

Qu'est-ce qu'un nuage ?

Ce à quoi l'on peut le plus probablement comparer un nuage, c'est à une éponge dans une baignoire. Tant qu'elle est sèche, l'éponge est légère comme une plume, mais à mesure qu'elle se gorge d'eau, elle devient de plus en plus lourde. Une fois saturée de liquide, elle se met à goutter. Il en va de même des nuages, qui sont des amas de minuscules gouttelettes d'eau ou cristaux de glace en suspension dans l'atmosphère. Quand un nuage est trop lourd, il se met à pleuvoir ou à neiger, selon les conditions météorologiques.

MAIS COMMENT LES NUAGES SE FORMENT-ILS AU JUSTE ?

Le principal responsable est, comme presque toujours dans la météo, le soleil. Quand il brille, l'eau des océans, des lacs, des cours d'eau et des flaques s'évapore : les particules d'eau à l'état gazeux, la vapeur d'eau, s'élèvent dans le ciel avec l'air chaud. Plus il fait chaud, plus l'air contient de particules d'eau. À des altitudes élevées, l'air est bien sûr plus froid et absorbe moins de particules d'eau. Ces dernières sont toujours plus à l'étroit jusqu'à ce que l'air, comme le disent les météorologues, soit « saturé ». C'est à ce moment précis que se forment les nuages. La vapeur d'eau se condense autour

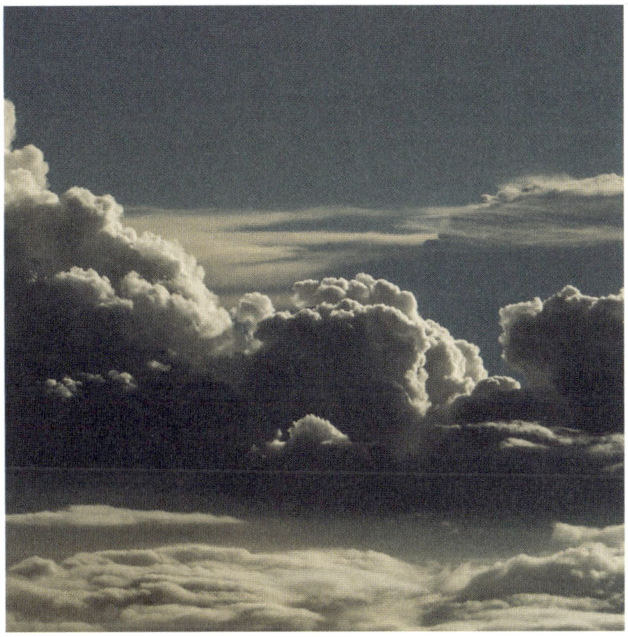

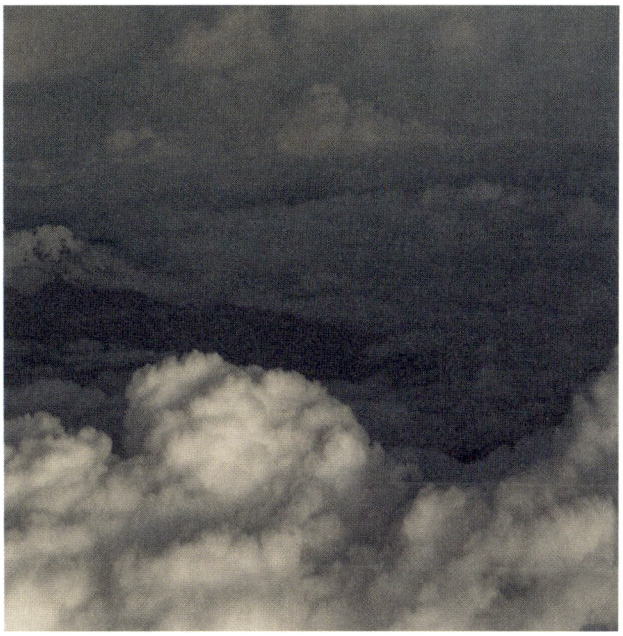

and this is the precise moment at which clouds come into existence. The water vapour crystallises around small aerosol particles—i.e. the tiniest particles of dust—and forms droplets. At a given point, the cloud becomes too heavy to keep these droplets and starts to rain.

Not each and every cloud will turn into rain: sometimes water vapour stops climbing into the air and the cloud continues on its peaceful course through the skies; sometimes, it simply dissolves. We weather experts talk about the water cycle: water evaporates, clouds are formed; the clouds turn to rain and this rainwater returns to the earth, from where it will once again evaporate.

Wolken entstehen. Dabei kondensiert der Wasserdampf an sogenannten Aerosolpartikeln, also kleinsten Staubteilchen – es entstehen wieder Tropfen. Bis die Wolke irgendwann zu schwer wird und die Wassertröpfchen nicht mehr an sich halten kann – es fängt an zu regnen.

Natürlich muss es nicht aus jeder Wolke regnen. Manchmal fehlt es am notwendigen Nachschub aus Wasserdampf, und die Wolke zieht einfach friedlich am Himmel entlang – oder löst sich gar wieder auf. Wir Wetterexperten sprechen von einem Wasserkreislauf: Wasser verdunstet, es entstehen Wolken. Die Wolken regnen sich aus, das Regenwasser kehrt zurück zum Boden – und verdunstet irgendwann erneut.

de particules aérosol – micropoussières – pour redevenir des gouttes d'eau. Quand le nuage finit par être trop lourd et ne peut plus retenir les gouttelettes d'eau, il pleut.

Cependant, chaque nuage n'engendre pas forcément de la pluie. S'il ne se forme pas suffisamment de vapeur d'eau, le nuage poursuivra sa route tranquillement dans le ciel, à moins qu'il ne se dissipe. Les météorologues parlent de cycle de l'eau : les nuages déversent leur eau de pluie, qui retourne au sol pour s'évaporer de nouveau un jour ou l'autre.

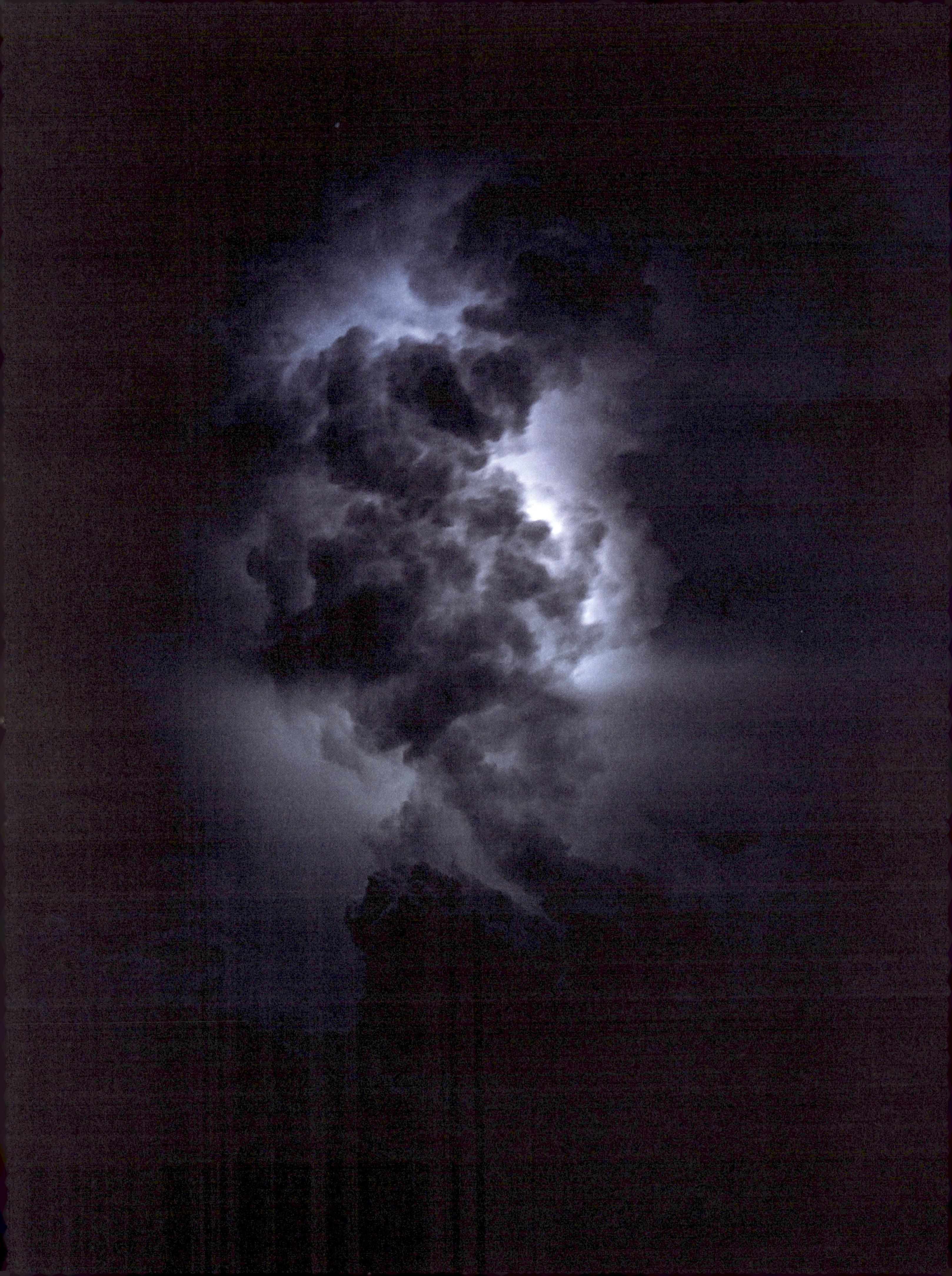

Stranger Things...?

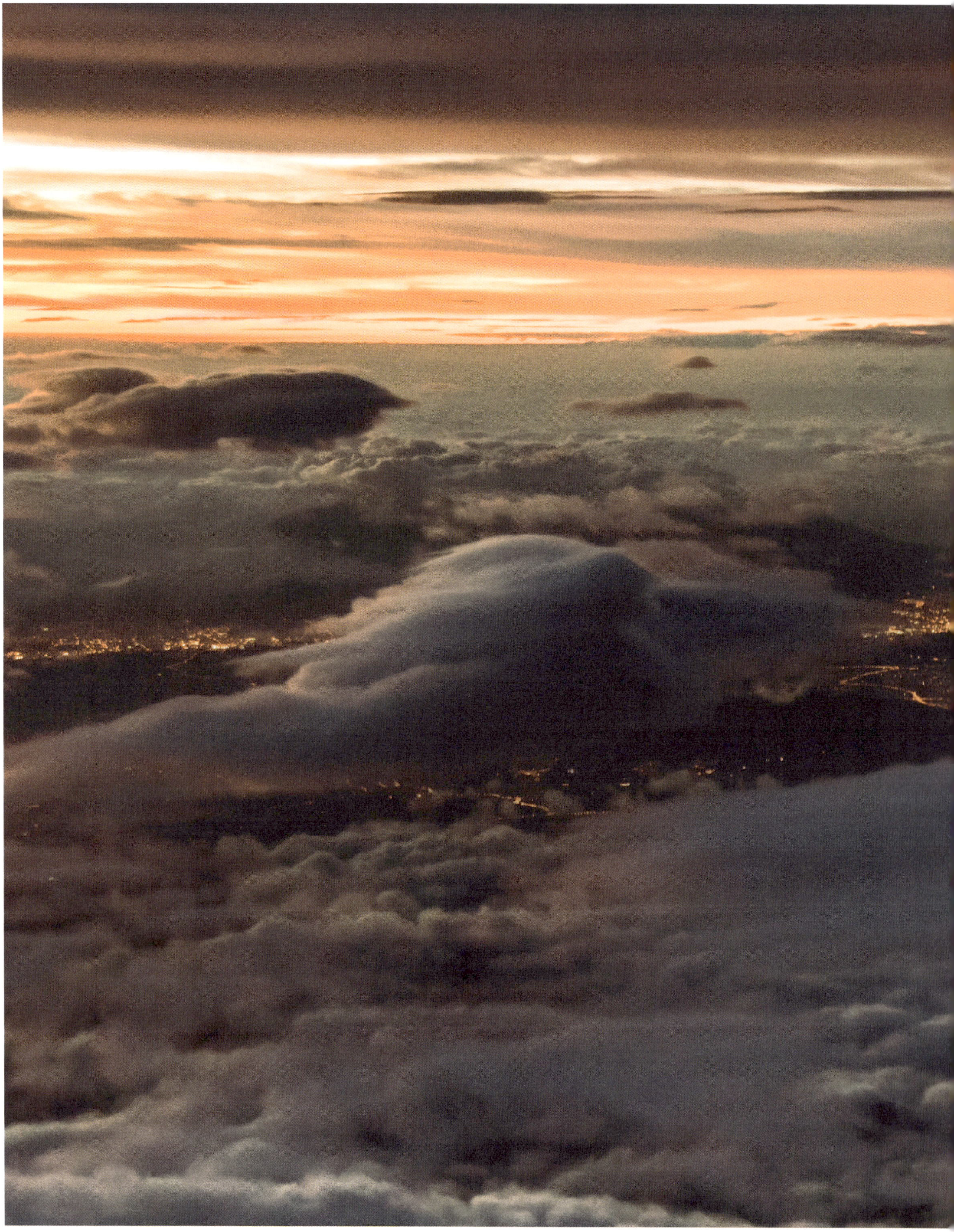

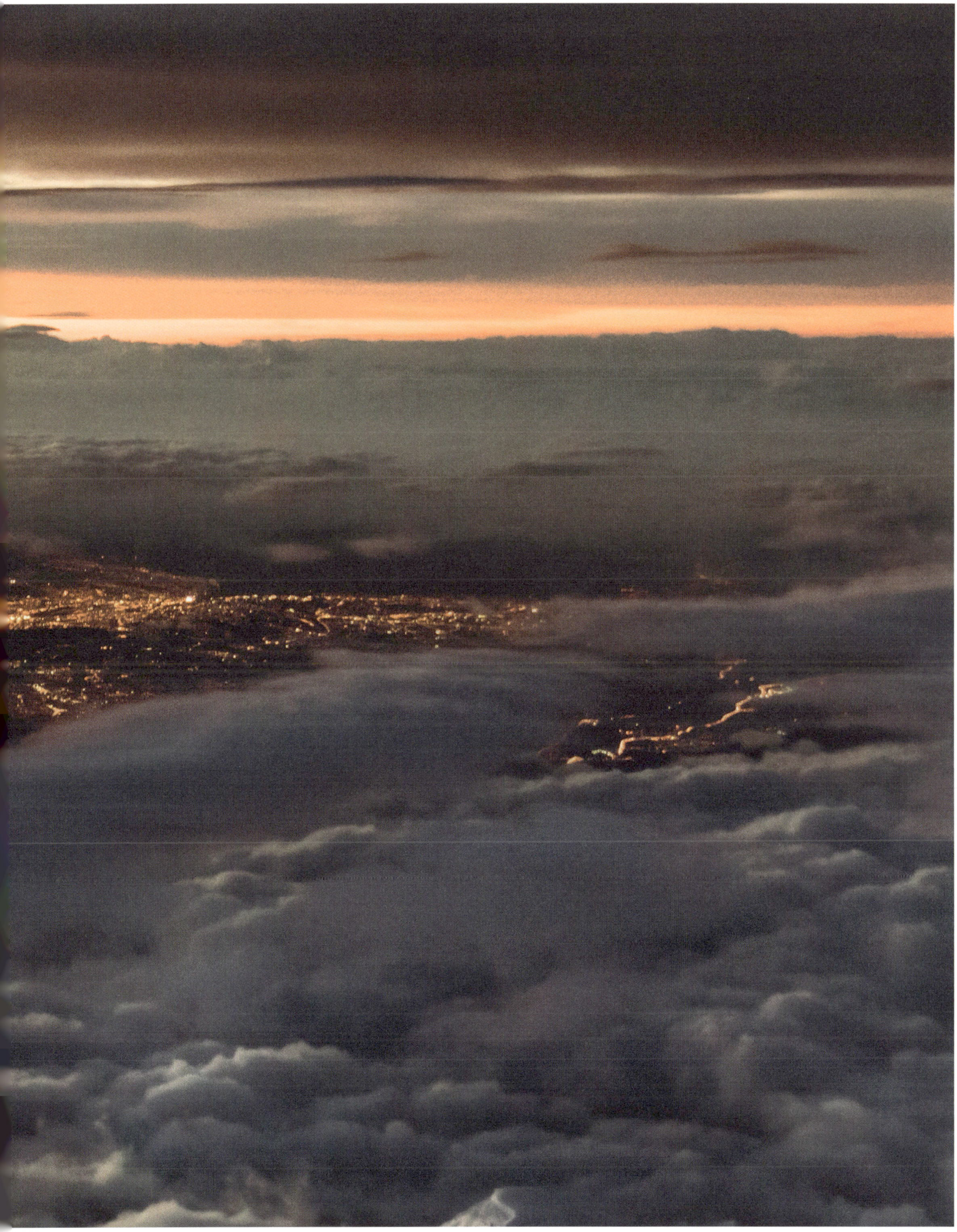

What kinds of cloud are there?

A good way to imagine clouds is to think about fog—which is actually nothing more than a cloud at ground level. Along with cumulus types, fog is one of the low-level clouds. They tend to consist entirely of water droplets and are generally found between altitudes of zero and 6,500 feet.

Slightly higher, at between 10,000 and 23,000 feet up, the middle layer of clouds is found. Named with the "alto-" prefix, they are mixed clouds, consisting both of water and ice.

Very high level clouds are called cirrus and are found at heights of between 16,500 and 43,000 feet; made entirely of ice crystals, these types of clouds don't bring any rain.

In addition to these three main kinds, meteorologists also talk about towering vertical clouds. These are the clouds which bring storms.

Each genus of clouds is classified into countless species and varieties, some of which have amusing or strange names: there is vertebratus, for instance, or translucidus. The first cloud atlas was published by the World Meteorological Organization (WMO for short), and to this day, any and every type of cloud can be found in it. In 2017, it was updated and made available online, too.

Welche Wolkenarten gibt es?

Am einfachsten lässt sich eine Wolke mit Nebel erklären. Denn Nebel ist nichts anderes als eine Wolke, die auf dem Boden aufliegt. Sie gehört damit – zusammen mit beispielsweise den Kumuluswolken – zu den sogenannten „tiefen Wolken“. Diese bestehen meist nur aus Wassertröpfchen und sind zwischen Erdboden und einer Höhe von zwei Kilometern zu finden.

Etwas höher liegen die „mittelhohen Wolken“. Sie tragen die meteorologische Vorsilbe Alto- und befinden sich zwischen zwei und sieben Kilometern Höhe. Es sind Mischwolken, teils aus Wasser, teils aus Eis bestehend.

Ganz oben wiederum schwirren die „hohen Wolken“ umher, die Cirren. Sie weilen zwischen fünf und 13 Kilometern Höhe und bestehen durchweg aus Eiskristallen. Diese Wolken bringen keinen Regen.

Die vierte Wolkenfamilie, die wir Wetterfrösche unterscheiden, sind Wolken mit großer vertikaler Erstreckung – die Gewitterwolken.

Jede dieser Familien hat unzählige Unterarten und -gattungen mit zum Teil witzigen und seltsamen Namen. Vertebratus zum Beispiel. Oder Translucidus. Die World Meteorological Organization, kurz WMO, publizierte 1896 den ersten Wolkenatlas, in dem sich jede beliebige Wolke nachschlagen lässt. 2017 wurde dieser Atlas aktualisiert und steht nun auch online zur Verfügung.

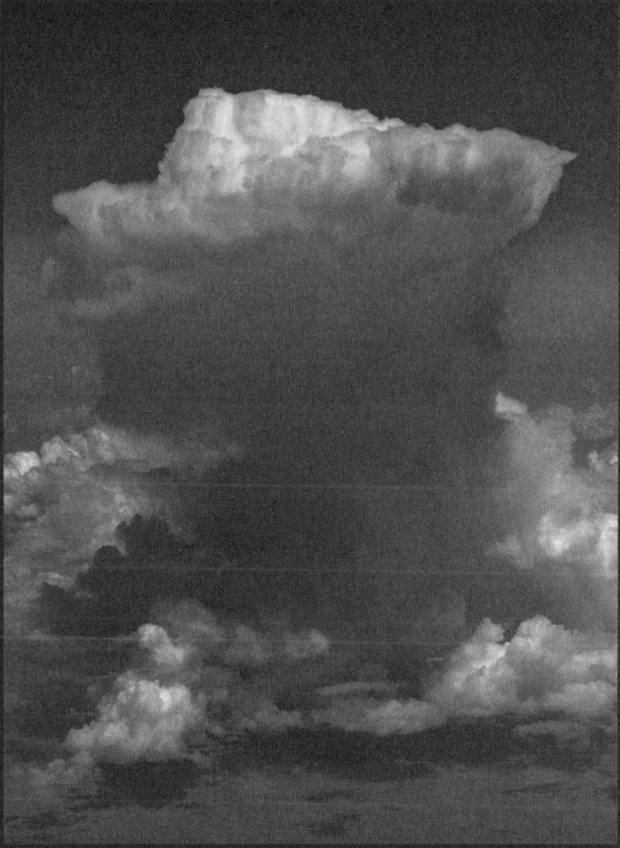

Quels sont les différents types de nuages ?

Pour expliquer les nuages, le plus simple est de parler du brouillard. Celui-ci n'est en effet rien d'autre qu'un nuage qui repose sur le sol et appartient, au même titre que les cumulus, à « l'étage inférieur ». Les nuages de cette catégorie sont en général constitués exclusivement de gouttelettes d'eau et se rencontrent entre le sol et 2 kilomètres d'altitude.

À une altitude comprise entre 2 et 7 kilomètres, les nuages de « l'étage moyen » portent des noms commençant par la syllabe « alto ». Il s'agit de nuages mixtes, à savoir composés d'eau et de glace.

À « l'étage supérieur », situé entre 5 et 13 kilomètres, évoluent les cirrus. Constitués exclusivement de cristaux de glace, ils ne donnent pas de pluie.

Enfin, les météorologues distinguent une quatrième famille : les nuages orageux, qui présentent une extension verticale importante.

Chacune de ces familles comprend une infinité de sous-catégories et de sous-genres, dont certains portent des noms aussi divertissants qu'étranges : vertebratus ou translucidus, pour ne citer qu'eux. L'Organisation météorologique mondiale (OMM) a publié en 1896 un *Atlas international des nuages* recensant tous les types de nuage. Depuis sa dernière mise à jour en 2017, cet atlas est également consultable en ligne.

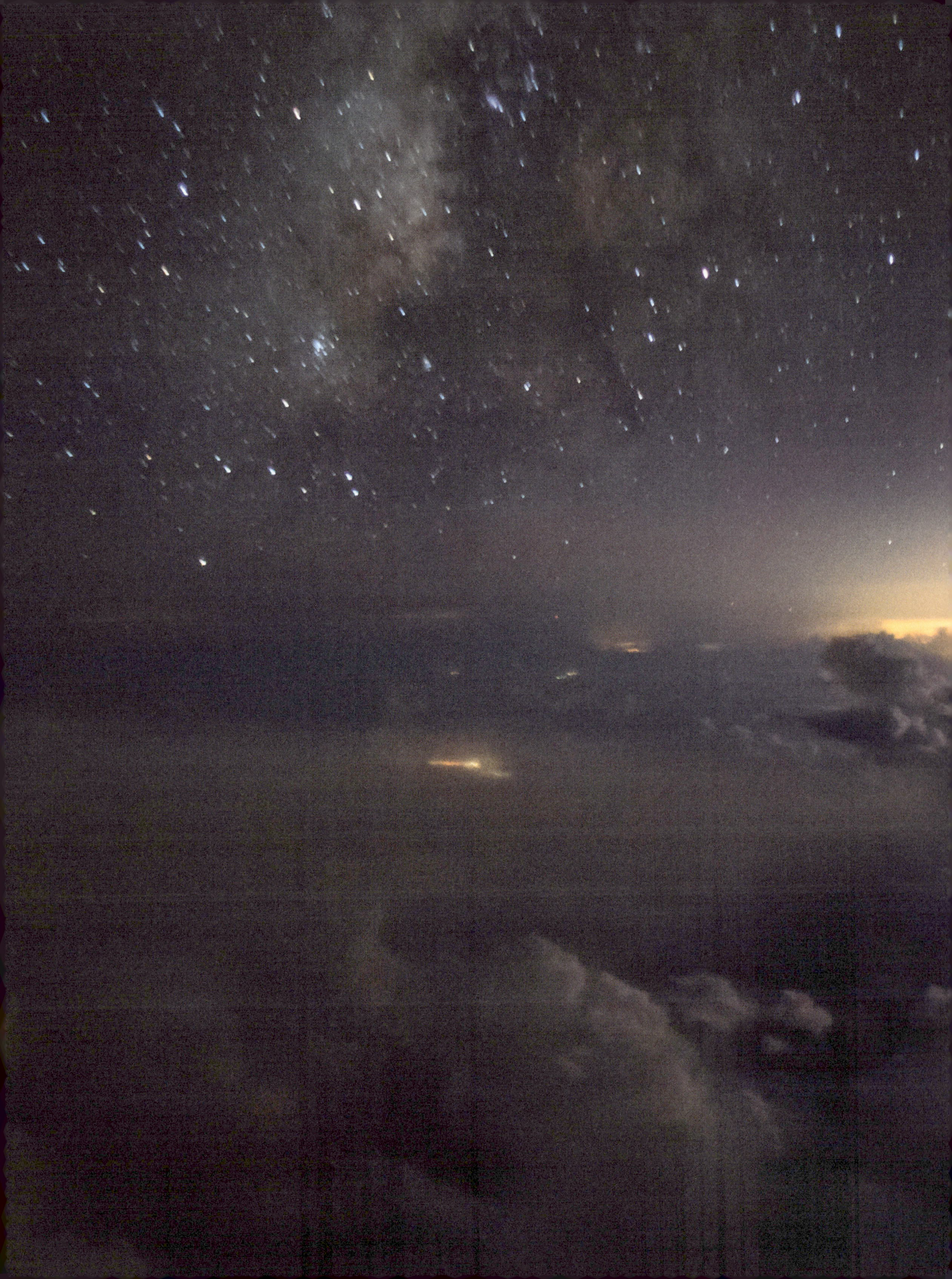

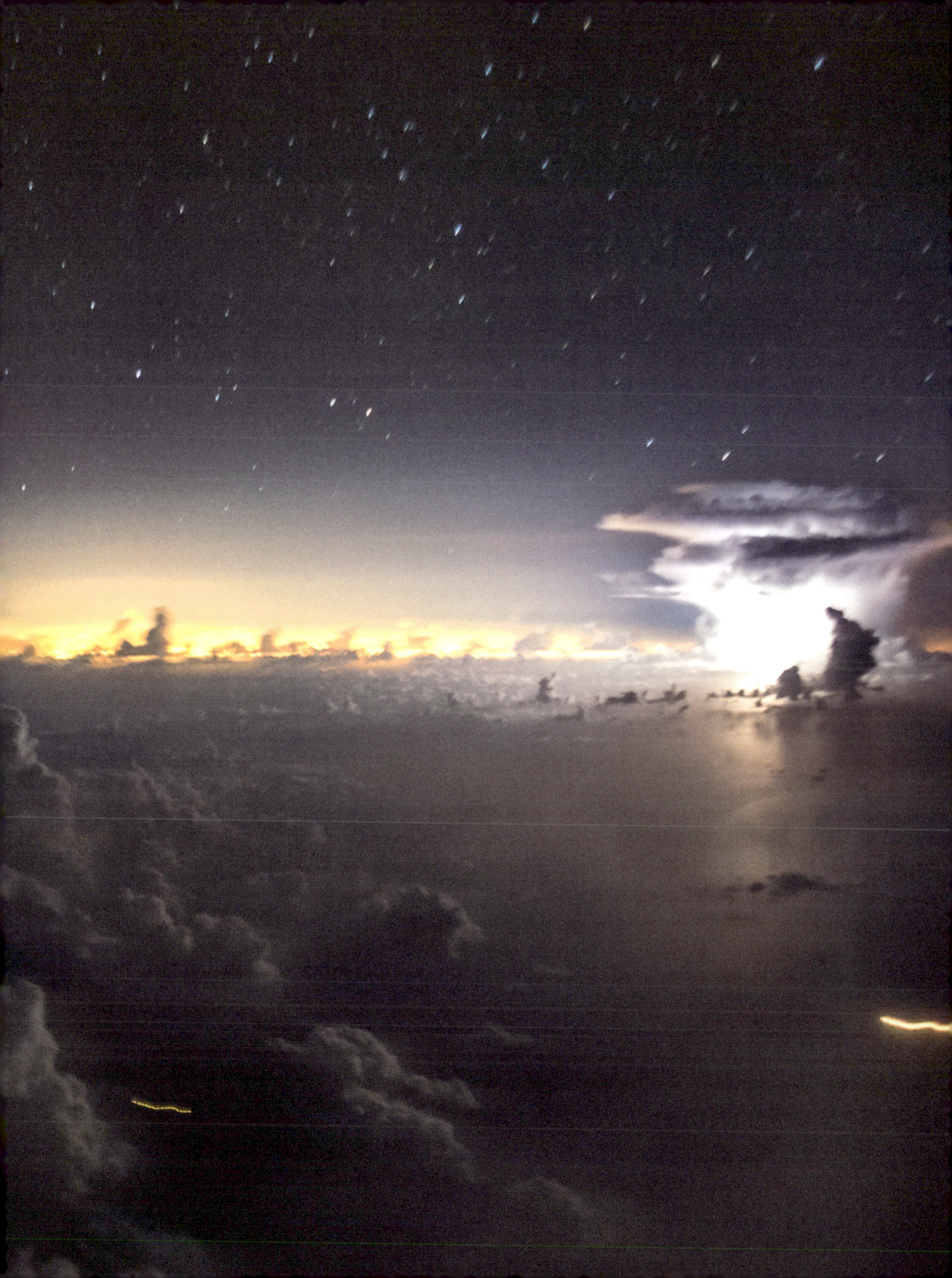

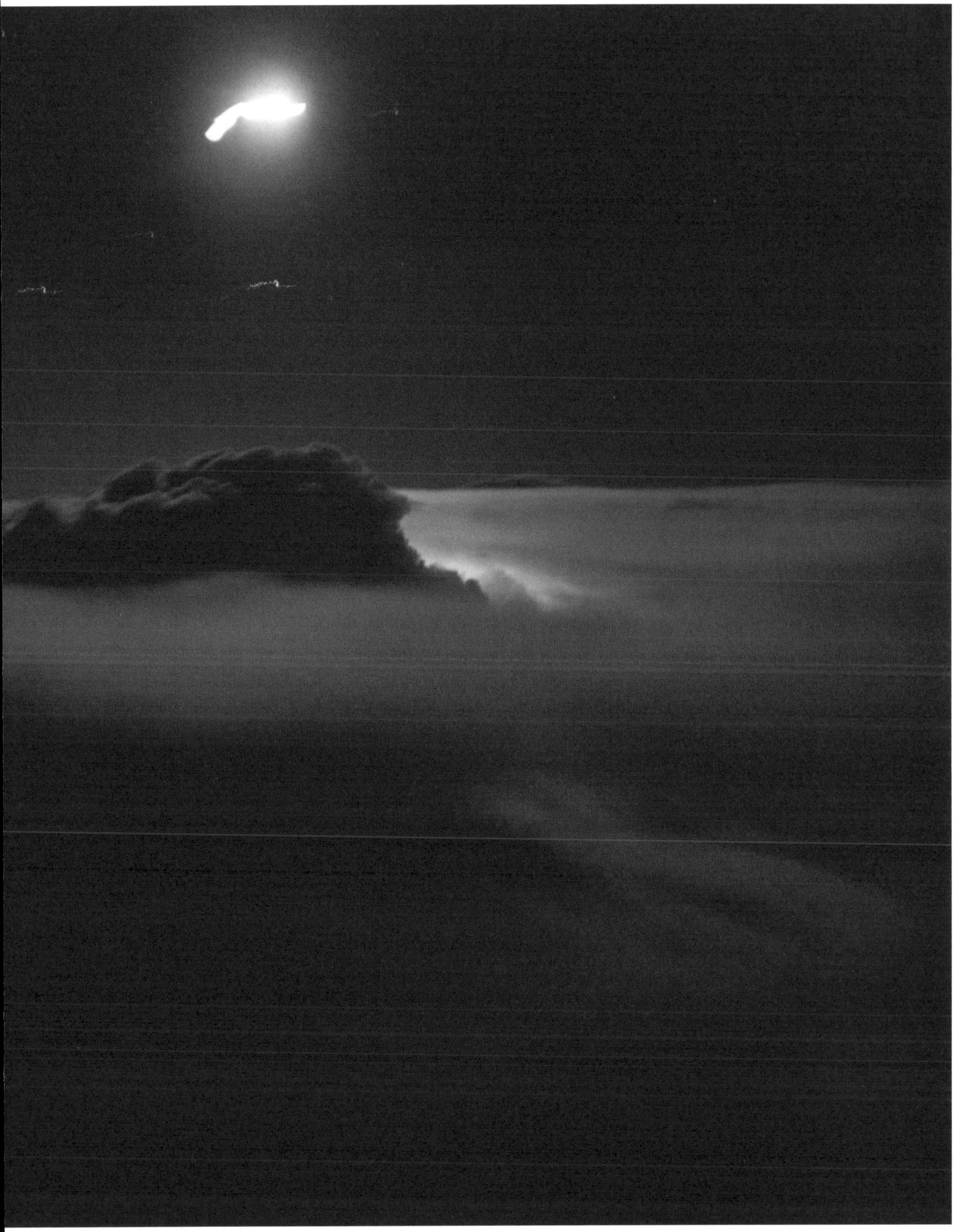

Why do clouds have different colours?

There are clouds that look like cotton-wool in bright white, then classic rain clouds with their slate grey; storm clouds can be so dark that they are almost black. Sometimes, one single cloud can appear in varying colours depending on where it is viewed from.

SO WHY DO CLOUDS TAKE ON SUCH A VARIETY OF SHADES?

We now know that all clouds are composed of tiny droplets of water or ice crystals. One thing that affects the colour we see is how many droplets a cloud contains and their size; the most important factor is, however, usually the point from which you are looking at a cloud, because the angle and intensity of sunrays is decisive. If a middle-height cloud is directly above us, for instance, it swallows up a lot of light from where we are standing: that makes it look grey. If that same cloud then moves on, it starts to look a lot lighter: the droplets of water in it are now acting like tiny mirrors, reflecting the sunlight in their thousands and millions. The result is that the cloud looks as white as the snow, especially around its edges. Clouds with large droplets of water absorb the most light, appearing dark,

Warum haben Wolken unterschiedliche Farben?

Schäfchenwolken sind weiß, Landregenwolken tauchen den Himmel in Grau, Gewitterwolken können dagegen fast schwarz aussehen. Zum Teil trägt sogar eine einzige Wolke – je nach Ort des Betrachters – verschiedene Farben.

WARUM ABER NEHMEN WOLKEN SO UNTERSCHIEDLICHE FARBTÖNE AN?

Wir wissen bereits: Alle Wolken bestehen aus winzig kleinen Wassertröpfchen oder Eiskristallen. Entscheidend für die Farbe ist zum einen, wie viele Tröpfchen sich in der Wolke befinden und welche Größe sie haben. Wichtiger ist aber häufiger der Standort des Betrachters – ausschlaggebend sind dabei der Winkel und die Intensität der Sonneneinstrahlung. Befindet sich eine mittelhohe Wolke genau über uns, verschluckt sie aus unserer Perspektive eine Menge Licht und erscheint dadurch grau. Ist die gleiche Wolke etwas später weitergezogen, wirkt sie schon viel heller. Die Wassertröpfchen in der Wolke wirken nun wie tausende kleine Spiegel und reflektieren das Sonnenlicht. Vor allem am Rand erscheint die Wolke nun fast schneeweiß. Vor allem großtropfige Wolken verschlucken extrem viel Licht und wirken dunkel

D'où viennent les colorations diverses des nuages ?

Les nuages moutonneux sont blancs, les nimbostratus contribuent aux ciels de grisaille, et les nuages orageux peuvent quant à eux paraître presque noirs. Un même nuage peut même, selon l'endroit d'où on l'observe, revêtir différentes couleurs.

POURQUOI LES NUAGES SE PARENT-ILS DE TEINTES AUSSI DIVERSES ?

Tous les nuages sont constitués de minuscules gouttelettes d'eau ou cristaux de glace. La couleur des nuages tient au nombre et à la taille de ces gouttelettes et, facteur souvent plus important, à l'angle d'observation et à l'intensité du rayonnement solaire. Si un nuage de l'étage moyen se trouve juste à notre aplomb, il absorbe une grande quantité de lumière incidente par rapport à nous et nous apparaît par conséquent gris. Si ce même nuage s'éloigne de nous ensuite, il semble déjà bien plus lumineux. Les gouttelettes d'eau qu'il contient font alors office d'autant de minuscules miroirs réfléchissant la lumière solaire. C'est principalement sur ses marges que le nuage semble alors presque blanc comme neige. Les nuages constitués de grosses gouttes d'eau absorbent d'importantes quantités de lumière

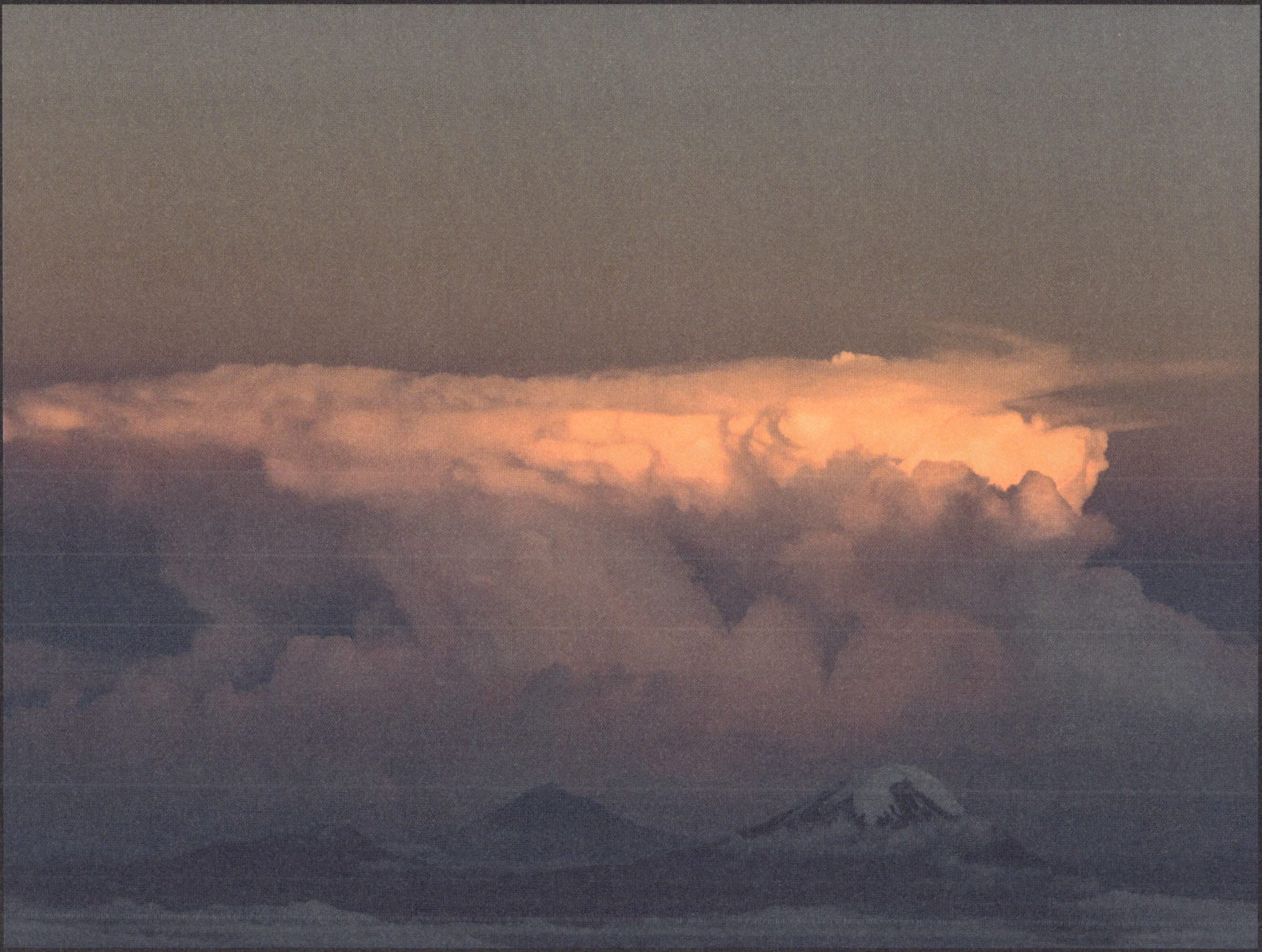

almost black; moreover, clouds often fall under the shadow of other clouds, which makes them look even darker.

A particularly good way to see this is to take off in an aeroplane while the weather is bad: the higher you get, the less distance rays of light have to travel through the cloud, and so the higher the cloud, the more light it receives—and the brighter it looks. Once you're above the clouds, they look like a sea of bright white candy floss.

bis schwarz. Häufig werden Wolken auch von den Schatten anderer Wolken getroffen, was sie noch düsterer macht.

Besonders gut kann man die Farbveränderung von Wolken beim Start eines Flugzeugs bei schlechtem Wetter beobachten. Je höher man steigt, desto kürzer ist die Strecke, die das Sonnenlicht durch die Wolken zurücklegen muss. Die Wolken weiter oben bekommen also viel mehr Licht ab und erscheinen heller. Nach Durchbrechen der Wolkendecke sieht man auf ein Meer aus strahlend weißen Zuckerwattewolken.

et semblent très sombres, voire noirs. Et quand des nuages projettent leur ombre sur d'autres nuages, ces derniers sont encore plus sombres.

C'est à bord d'un avion décollant par temps couvert que l'on peut vraiment bien observer ces variations de couleurs. Plus on prend de l'altitude, plus le trajet de la lumière solaire à travers les nuages est court. Ainsi, plus un nuage est en altitude, plus il reçoit de lumière et plus il semble clair. Une fois au-dessus de la couche nuageuse, on survole un océan de nuages cotonneux d'un blanc immaculé.

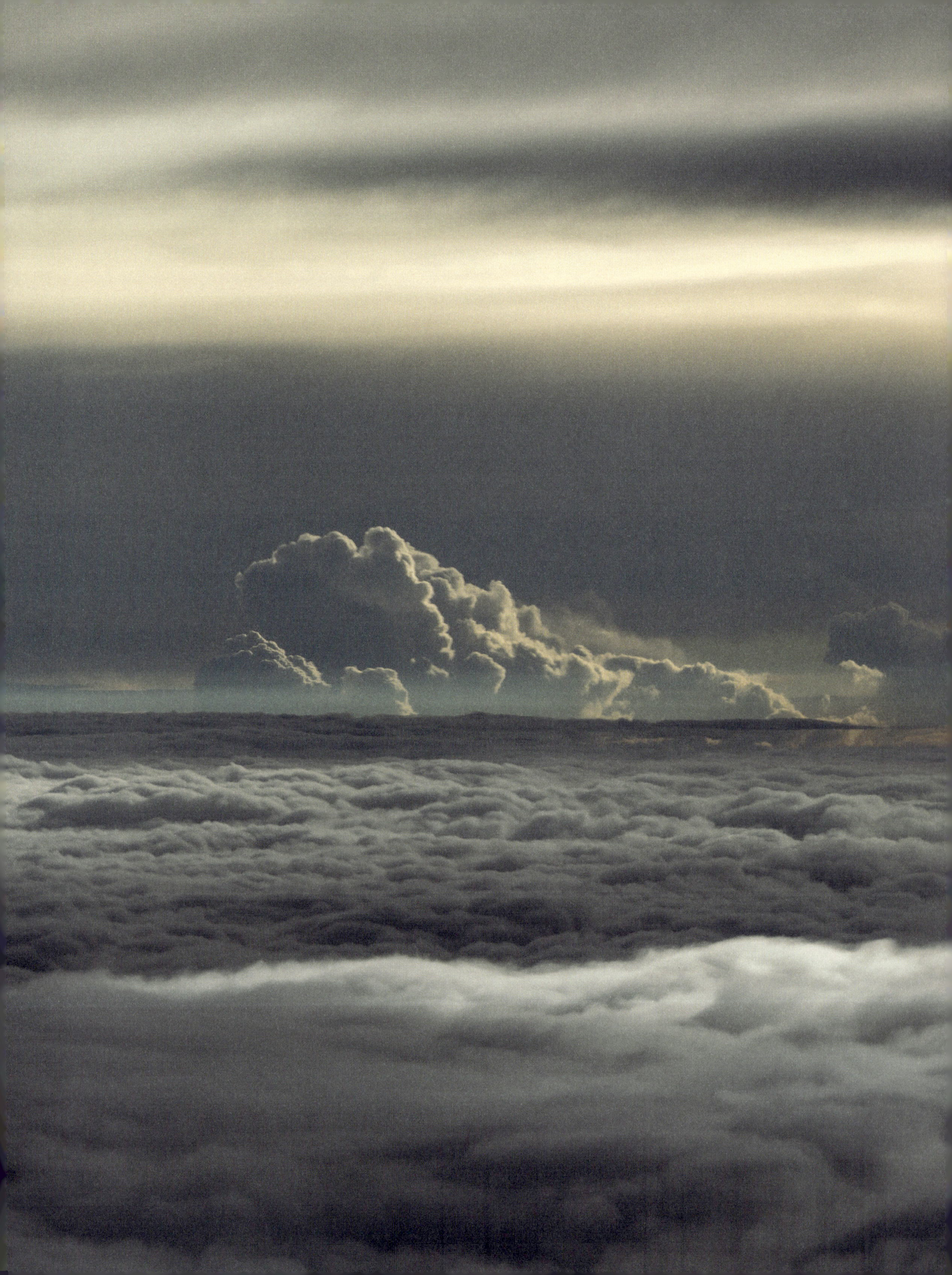

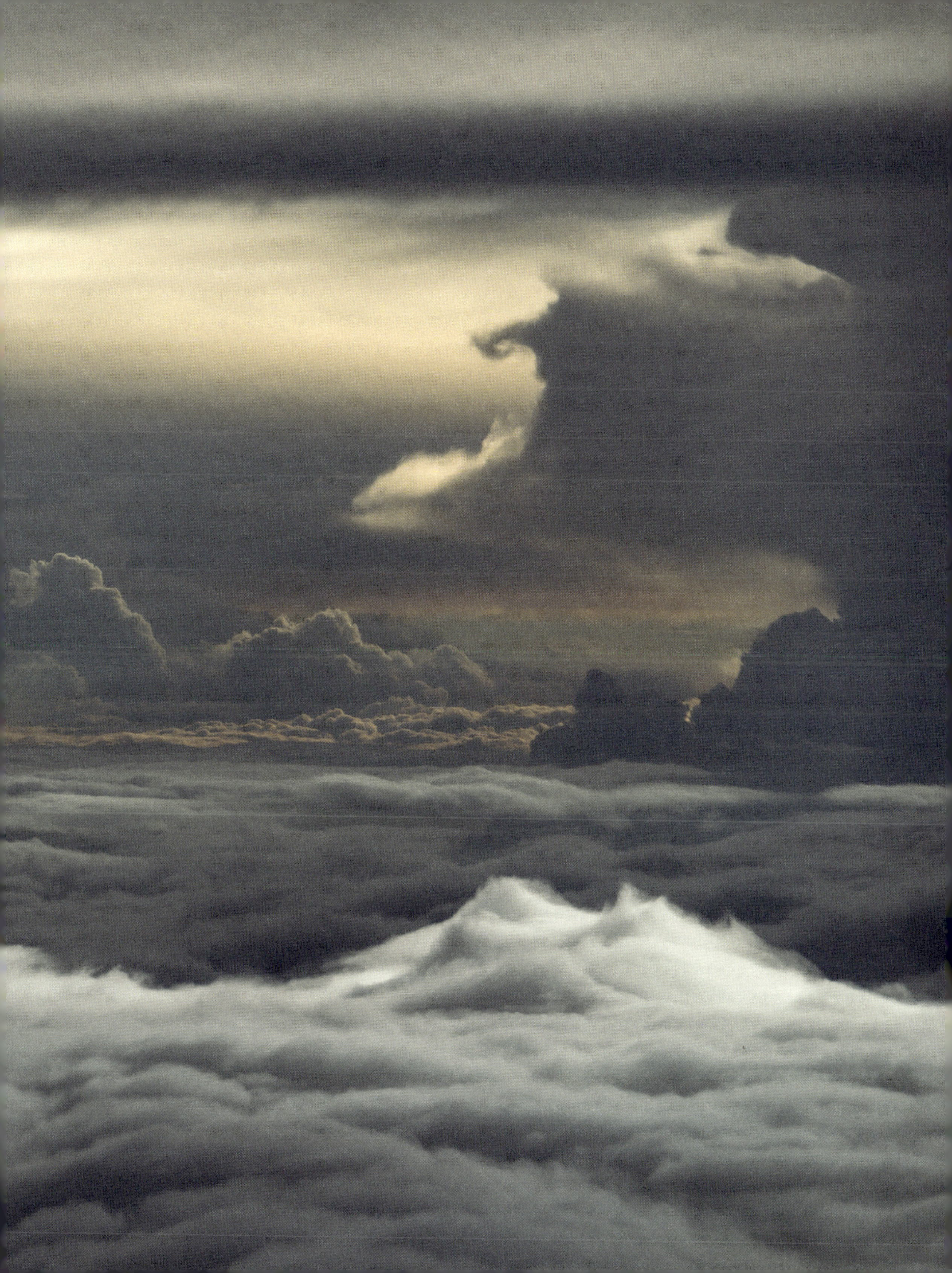

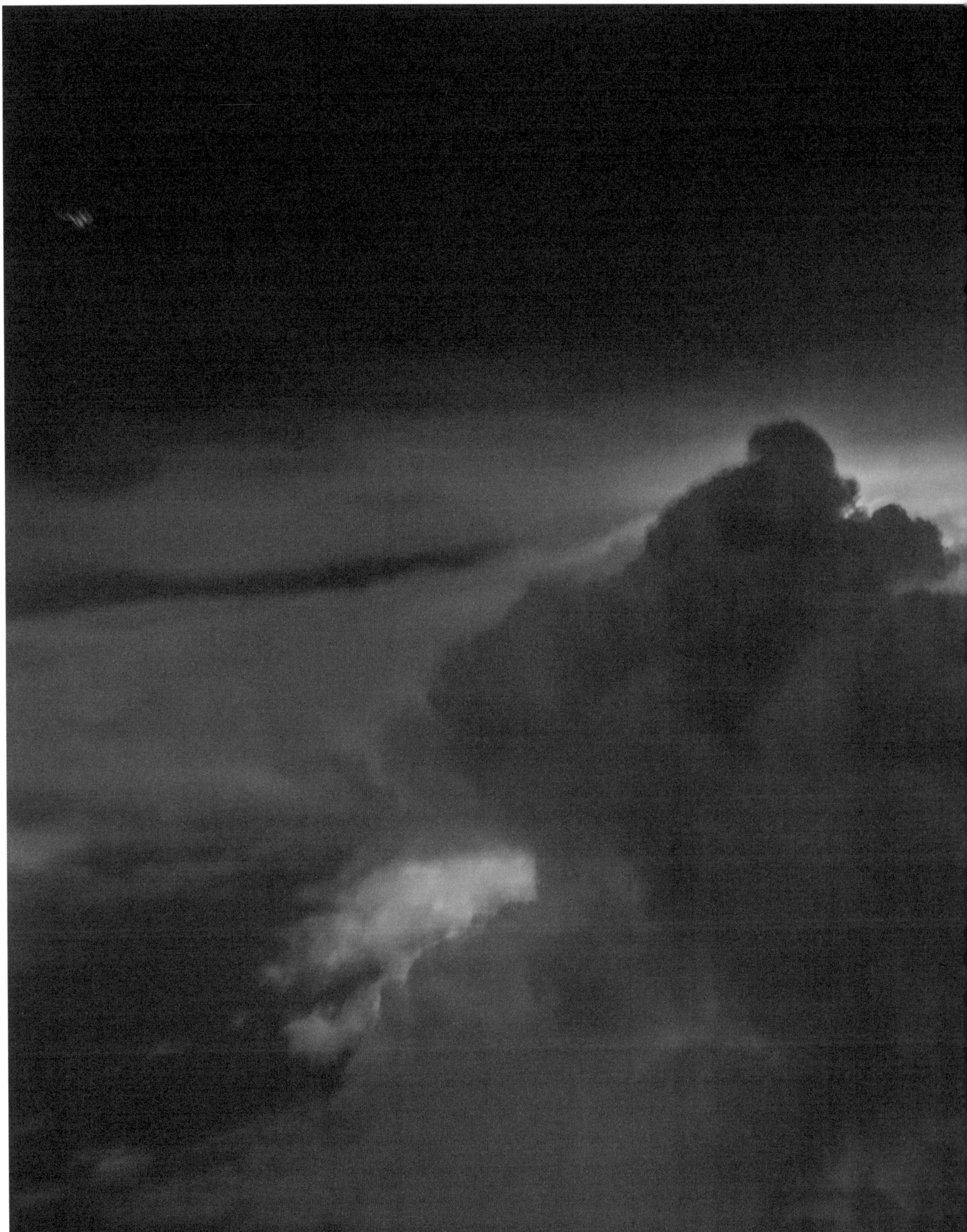

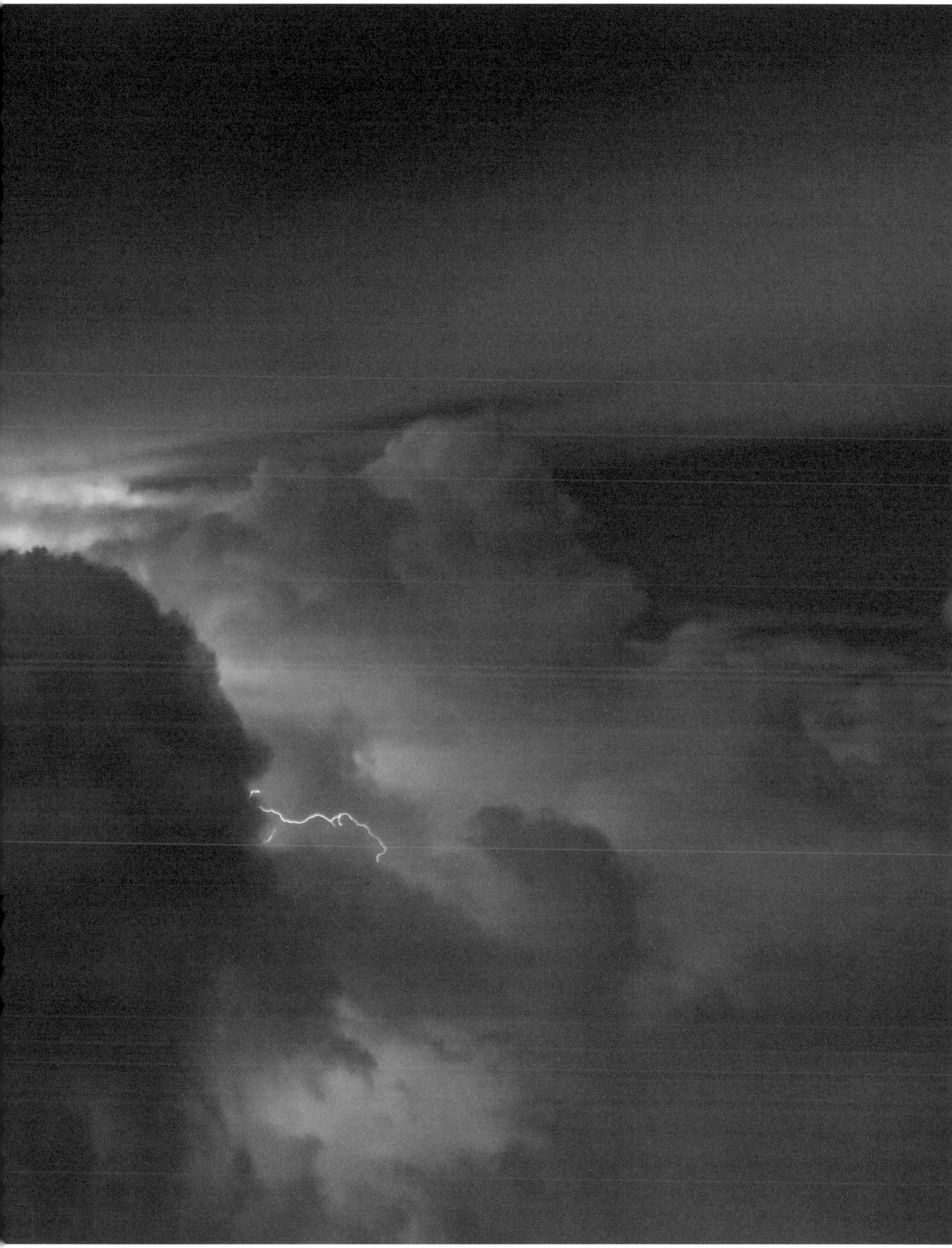

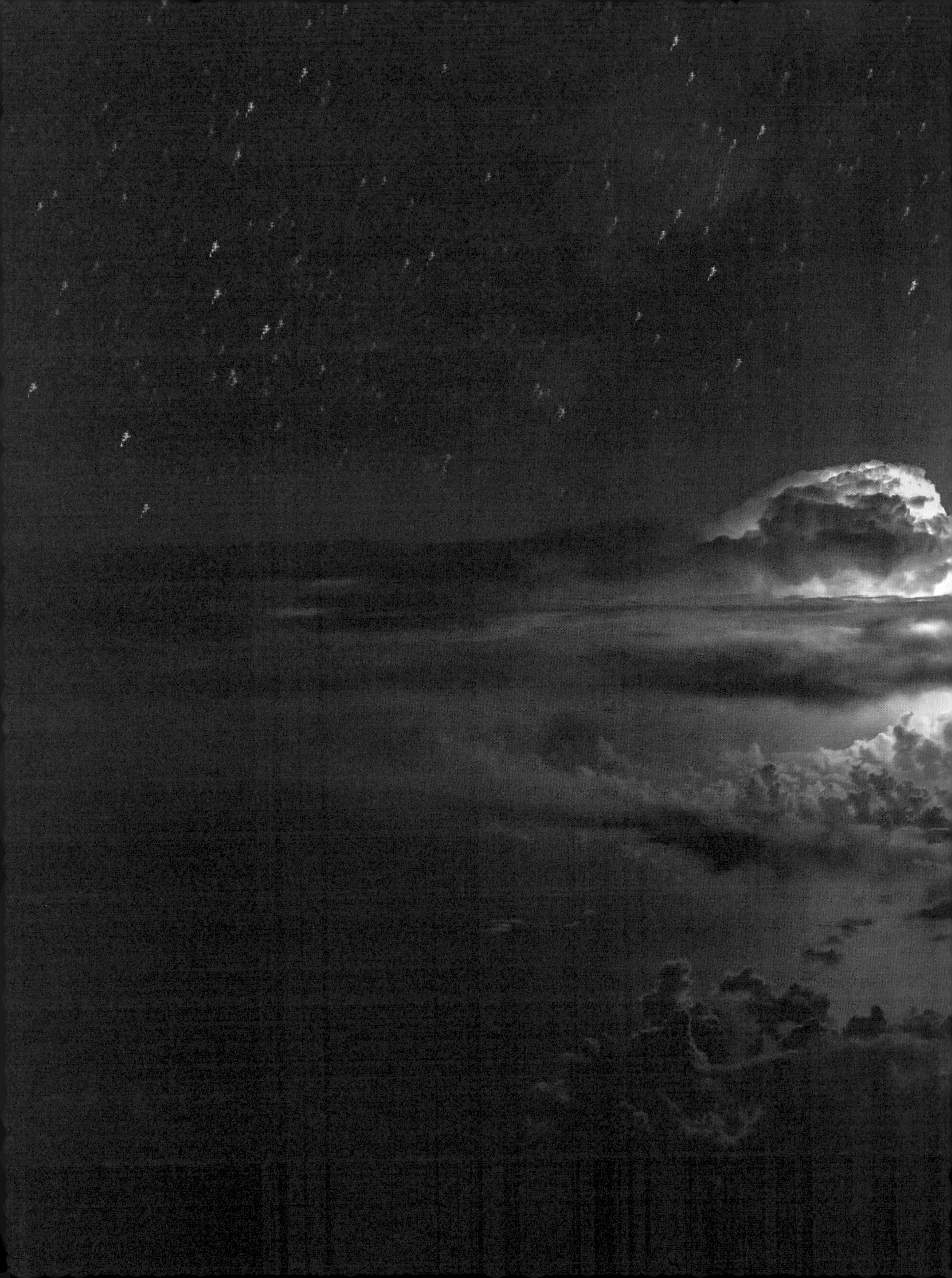

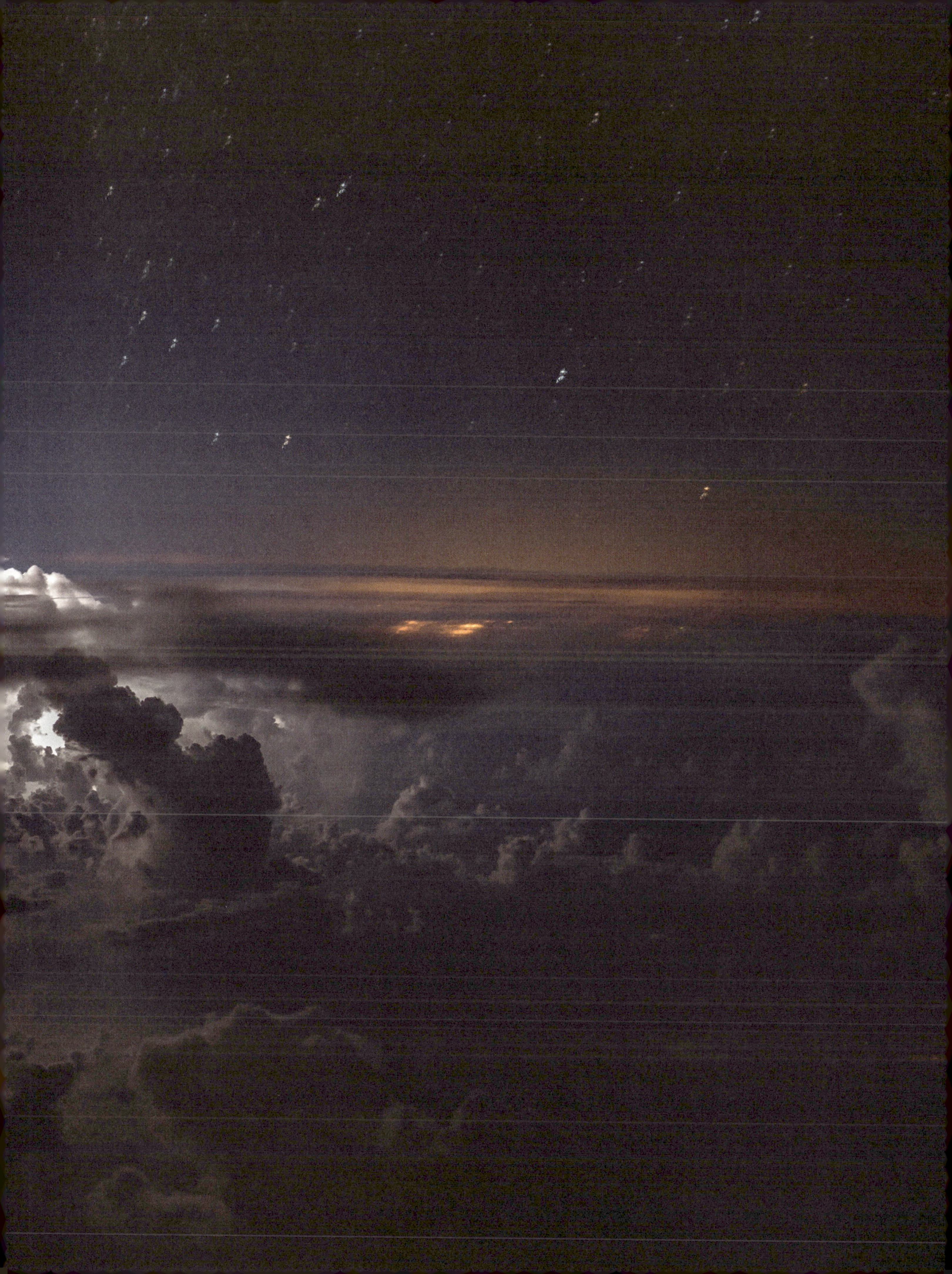

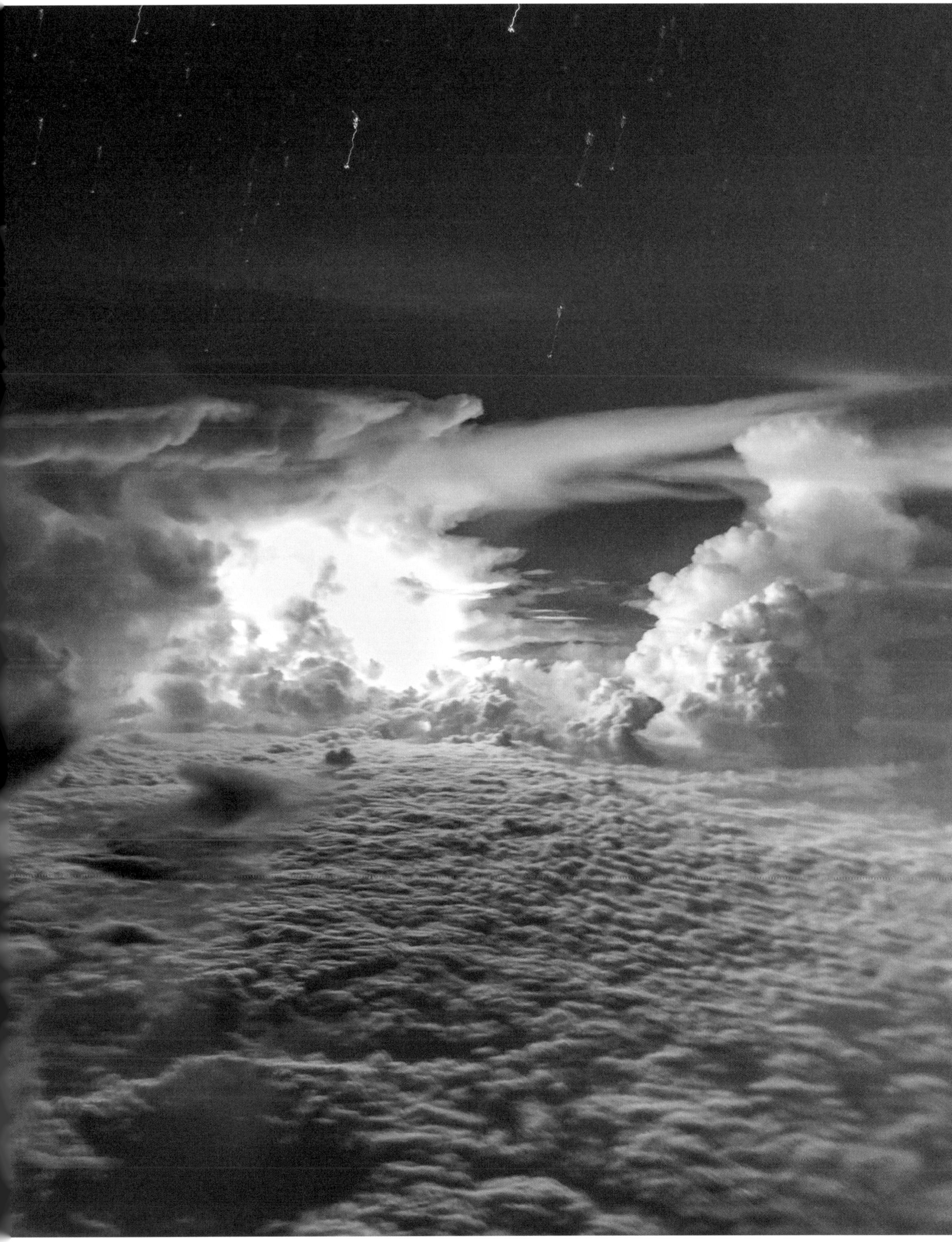

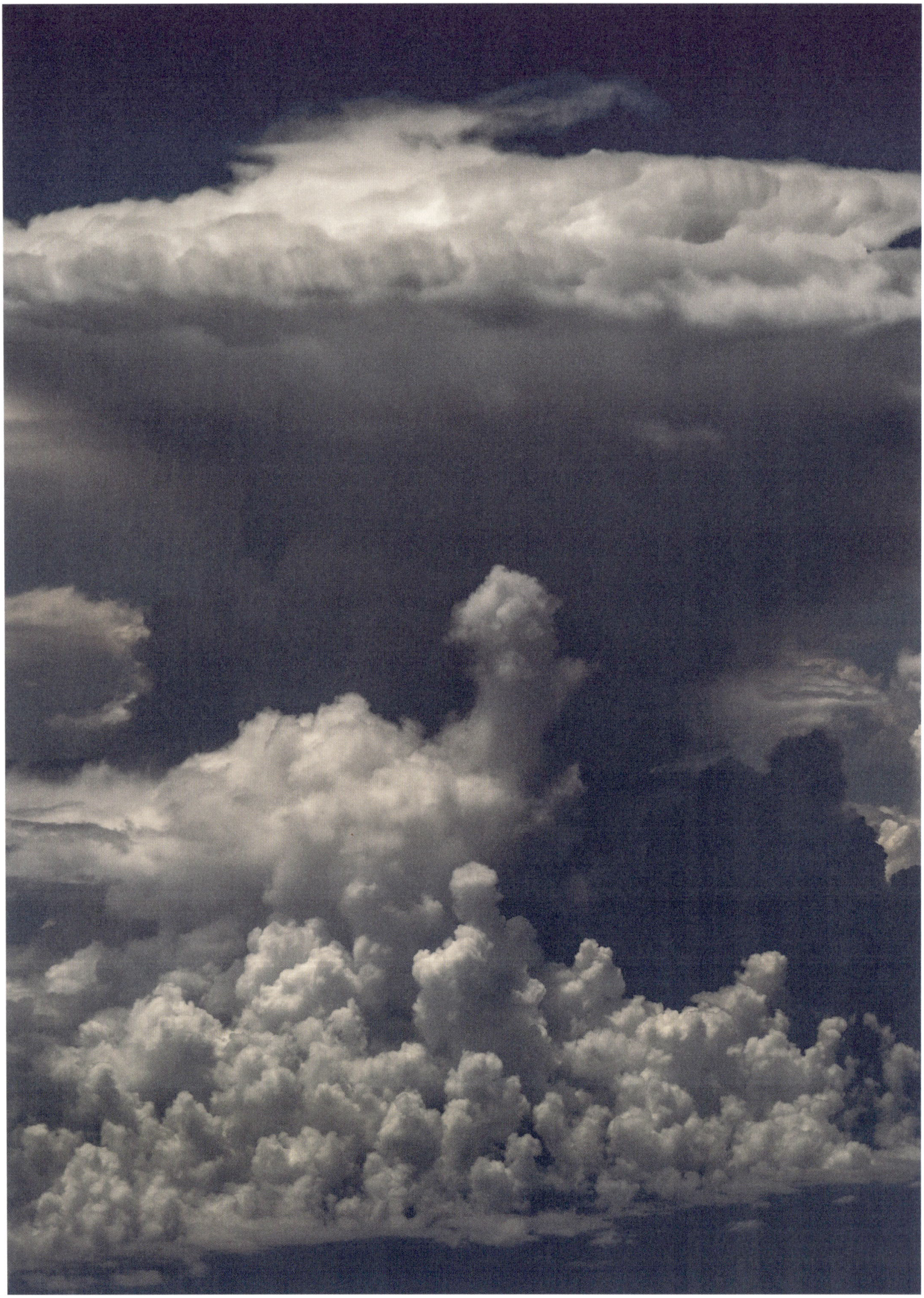

Cumulus. It's hard to realize they're just water #cumulonimbus #cumulus #weather #storm #lightning

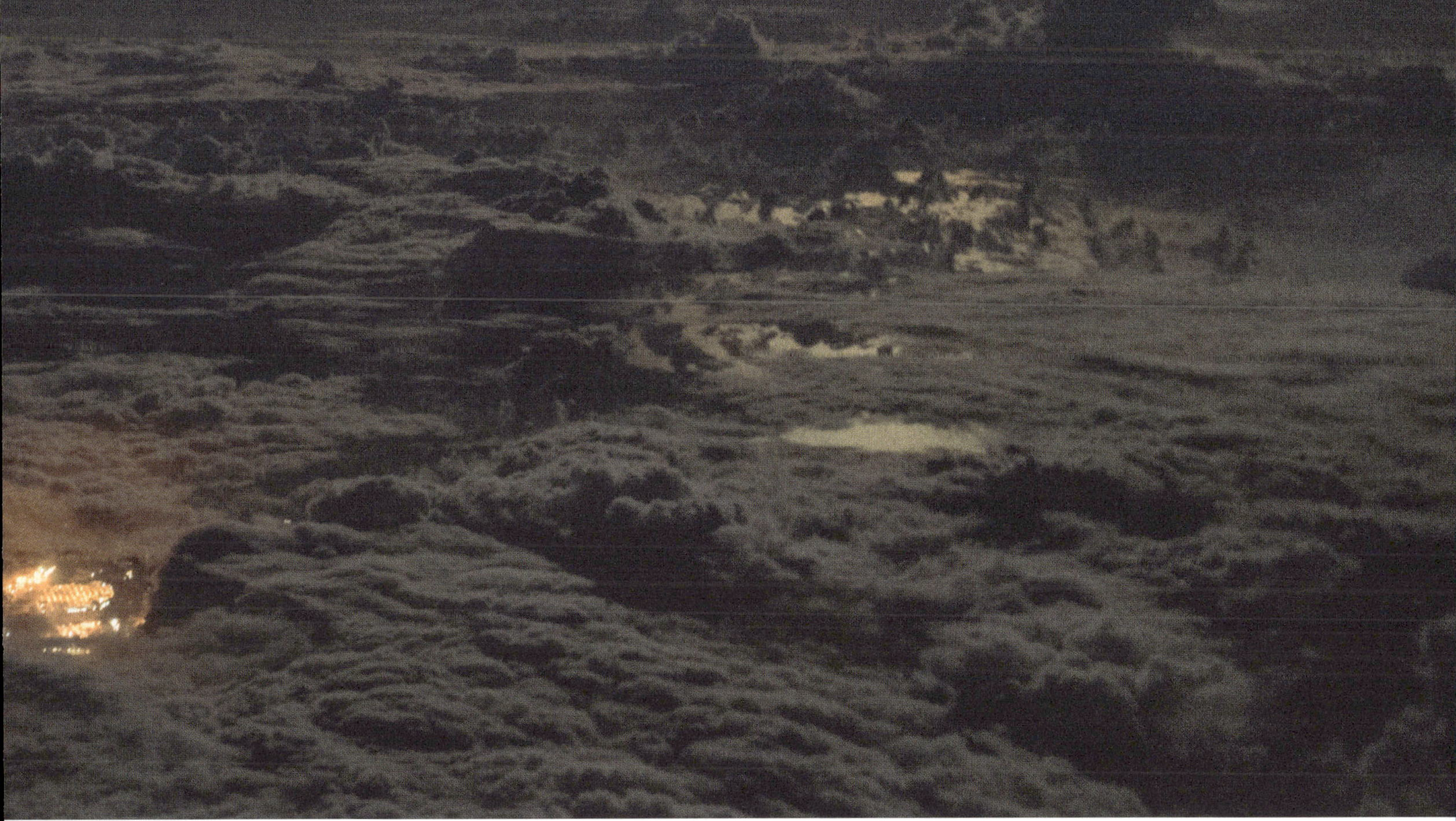

What planet is this?

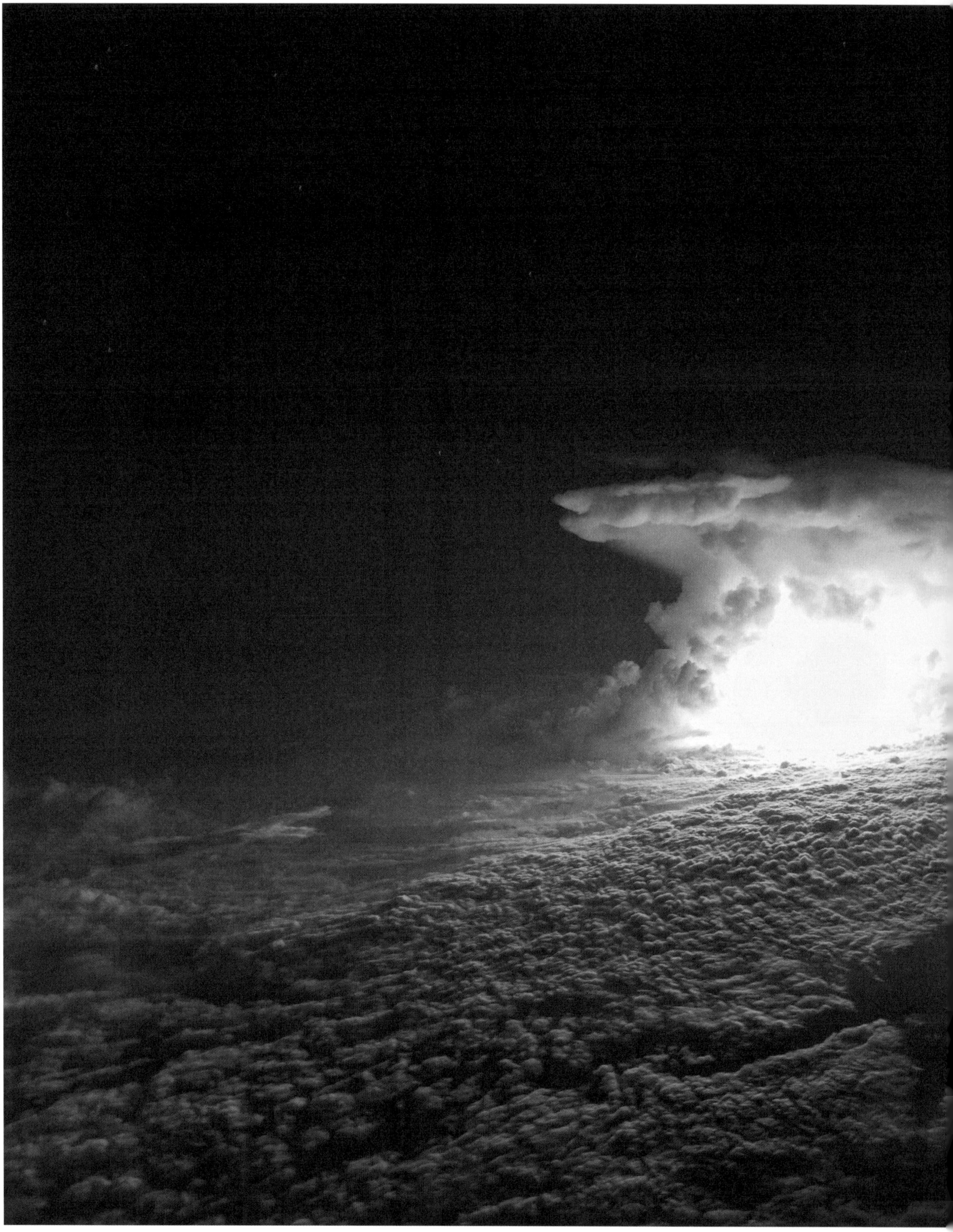

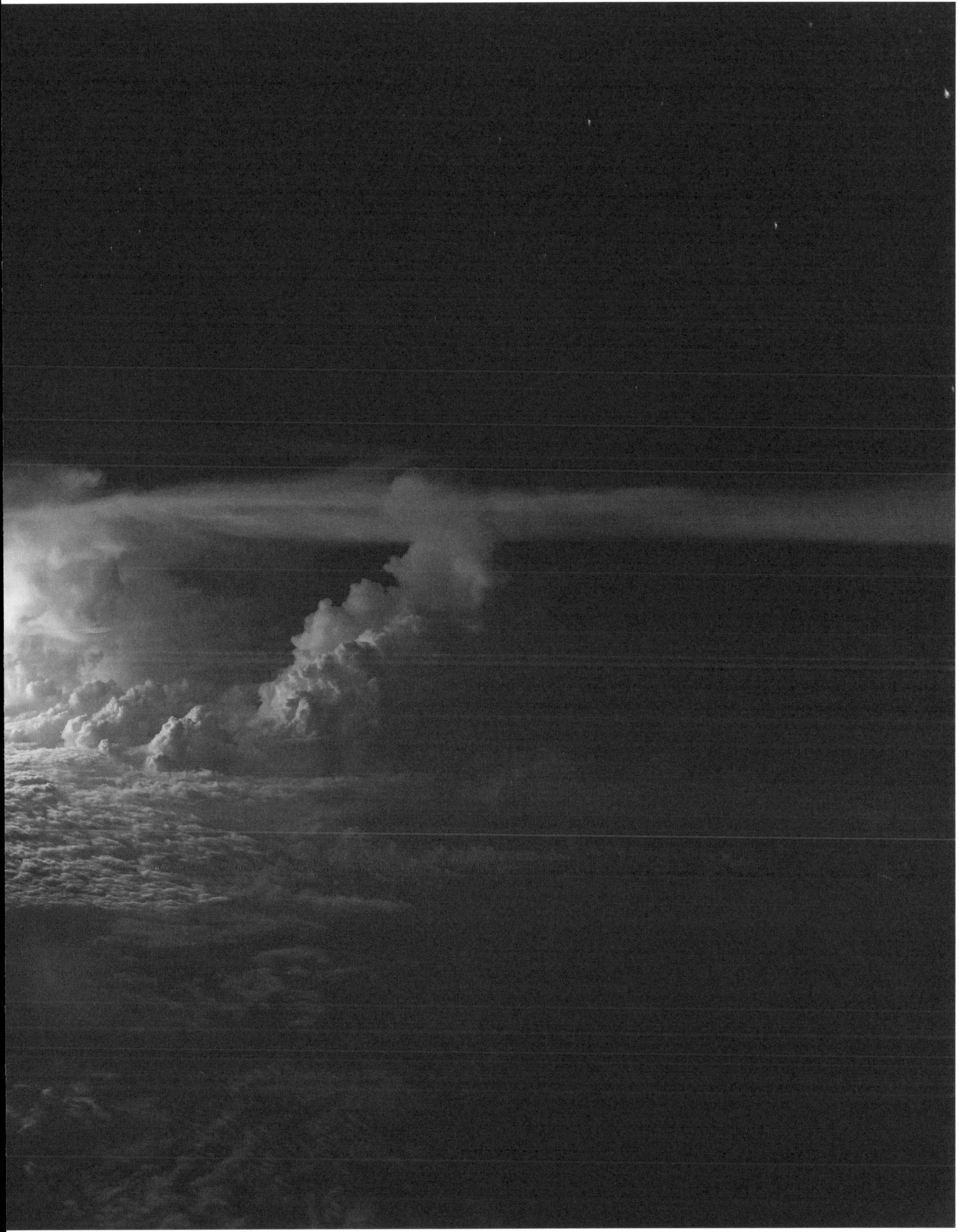

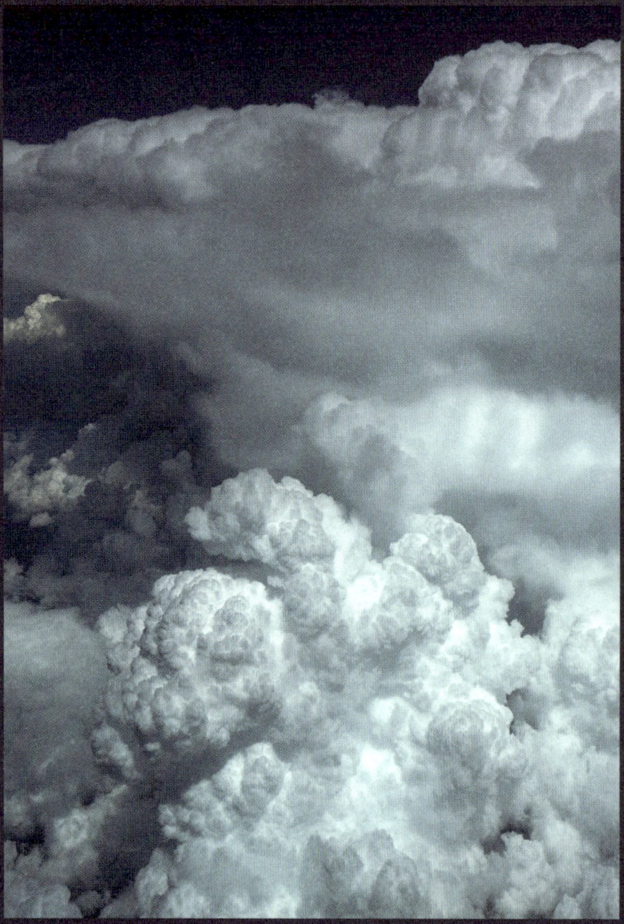

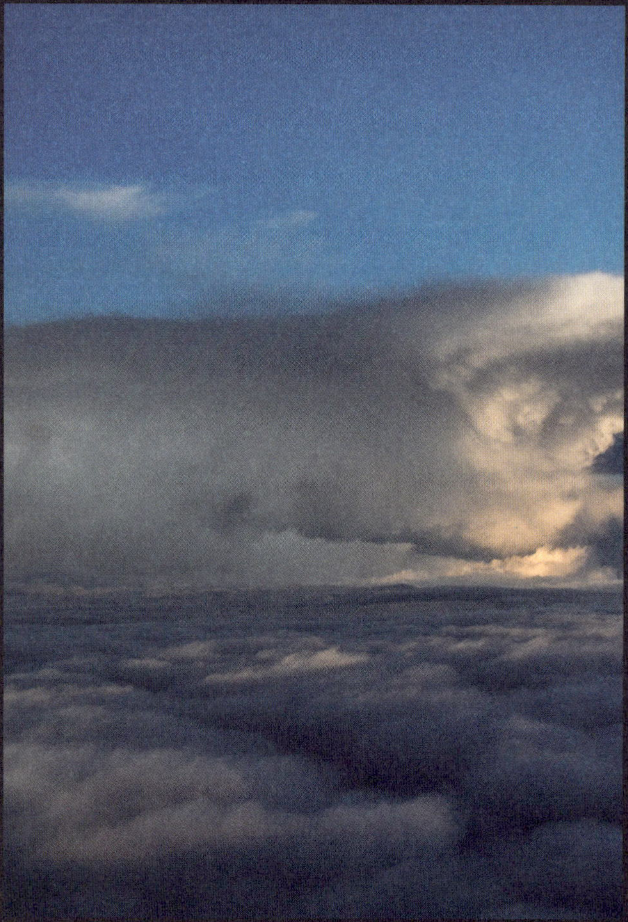

How is cloud coverage classified?

Around two thirds of the globe regularly experiences cloud cover. Oceans in which weather fronts regularly develop in temperate zones and the tropical rain belt around the equator have particularly frequent cloud cover; over deserts, by contrast and in the subtropical seas, clouds are a relatively rare occurrence. Meteorologists divide the sky up into eight slices—like in a pie chart—and report coverage in "oktas". If there is just one small cloud in the sky, that's 1/8 cloud coverage ("sunny"); 2/8 is "fair", 3/8 is "light cloud cover" and, taking us to the halfway point, 4/8, is "cloudy". 8/8 is complete cloud coverage under which no clear sky can be seen. The worst rating, however, is an entertainingly impossible 9/8, which means that you can't even see the cloudy sky—because you're inside a patch of fog.

Information about cloud coverage is especially important for pilots who need to navigate safely from one point to another. They have their own abbreviations such as NSC—'no significant clouds', or no cloud cover under 5000 feet—and OVC, or overcast (i.e. full cloud coverage), which is bad for flying. The worst possible case is reported as OBSC, however: obscured. That means that there is no point at which clear skies can be seen. Measurement is generally done by satellite cameras, which when linked up with ground-level sensors, can differentiate between high, middle, and low clouds. For professional aviators, amateur pilots, and mountaineers, the cloud base—the lowest point of a cloud—is an important item of information, too.

Was bedeutet Bedeckung?

Zwei Drittel unseres Globus sind regelmäßig mit Wolken bedeckt. Vor allem Gebiete mit Sturmtiefs über den Ozeanen der mittleren Breiten sowie der tropische Regengürtel um den Äquator sind sehr wolkenreich. Dagegen gibt es vor allem über den Wüsten und subtropischen Meeren relativ wenige Wolken. Meteorologen teilen den Himmel wie einen Kuchen in acht Teile. Wenn nur ein einziges kleines Wölkchen am Himmel auftaucht, spricht man von einer 1/8-Bewölkung, es ist „sonnig". So geht es weiter über „heiter" (2/8) und „leicht bewölkt" (3/8) bis zu „wolkig" (4/8) – der halbe Kuchen. 8/8 äußert sich natürlich durch eine flächendeckende Bedeckung, bei der der freie Himmel gar nicht mehr zu sehen ist – es geht allerdings noch schlimmer. 9/8 gibt es lustigerweise auch. Der Himmel ist dann gar nicht mehr erkennbar: Man steht im Nebel.

Für Piloten ist die Bewölkung besonders wichtig, um ihre Flugzeuge sicher von A nach B zu bekommen. Sie sprechen ihre eigene Sprache: NSC bedeutet *No significant Clouds* – keine Wolken unter 5000 Fuß. Schlecht für die Fliegerei ist OVC: *Overcast,* eine geschlossene Wolkendecke. Für den schlimmsten Fall aber steht das Kürzel OBSC: *Obscured.* Dabei ist der freie Himmel an keiner Stelle mehr sichtbar. Die Messungen dazu übernehmen meist Sensoren von Satellitenkameras. In Kombination mit Bodensensoren kann dann zwischen hohen, mittelhohen und tiefen Wolken unterschieden werden. Besonders in der Luftfahrt, Sportfliegerei und in Hochgebirgen ist auch die Wolkenuntergrenze wichtig – der tiefste Punkt einer Wolke: die Wolkenbasis.

Qu'entend-on par nébulosité ?

Deux tiers du globe sont régulièrement couverts de nuages. Certaines régions, tels les océans à fortes perturbations atmosphériques sous les latitudes moyennes et la ceinture de pluie tropicale qui enserre l'équateur, le sont tout particulièrement. En revanche, les déserts et les mers subtropicales le sont rarement. Les météorologues partagent la voûte céleste en huit, comme un gâteau. Si un petit nuage isolé traverse le ciel, on estime la nébulosité à 1/8e du ciel : le ciel est « presque entièrement dégagé ». Une nébulosité de 2/8e correspond à un ciel largement dégagé, de 3/8e à un ciel légèrement nuageux, de 4/8e à un ciel assez nuageux, et de 8/8e à un ciel couvert. Mais il y a pire : la nébulosité de 9/8e, qui ne permet pas de distinguer le ciel, autrement dit le brouillard.

La nébulosité revêt une importance particulière pour les pilotes, qui doivent conduire leurs avions à bon port. Dans leur jargon, NSC signifie *no significant clouds,* à savoir pas de nuages en-dessous de 5 000 pieds d'altitude. OVC *(overcast)*, c'est-à-dire couvert, constitue de mauvaises conditions de vol. L'abréviation OBSC, pour *obscured,* est synonyme des pires conditions météo : le ciel est entièrement occulté. La nébulosité est en général mesurée par les capteurs d'appareils photo satellitaires. En recoupant avec les mesures effectuées par des capteurs au sol, il est possible de distinguer entre les nuages de l'étage supérieur, ceux de l'étage moyen et ceux de l'étage inférieur. Pour les transports aériens et l'aviation sportive ainsi qu'en haute montagne, une autre donnée revêt elle aussi de l'importance : la base des nuages, c'est-à-dire leur point le plus bas.

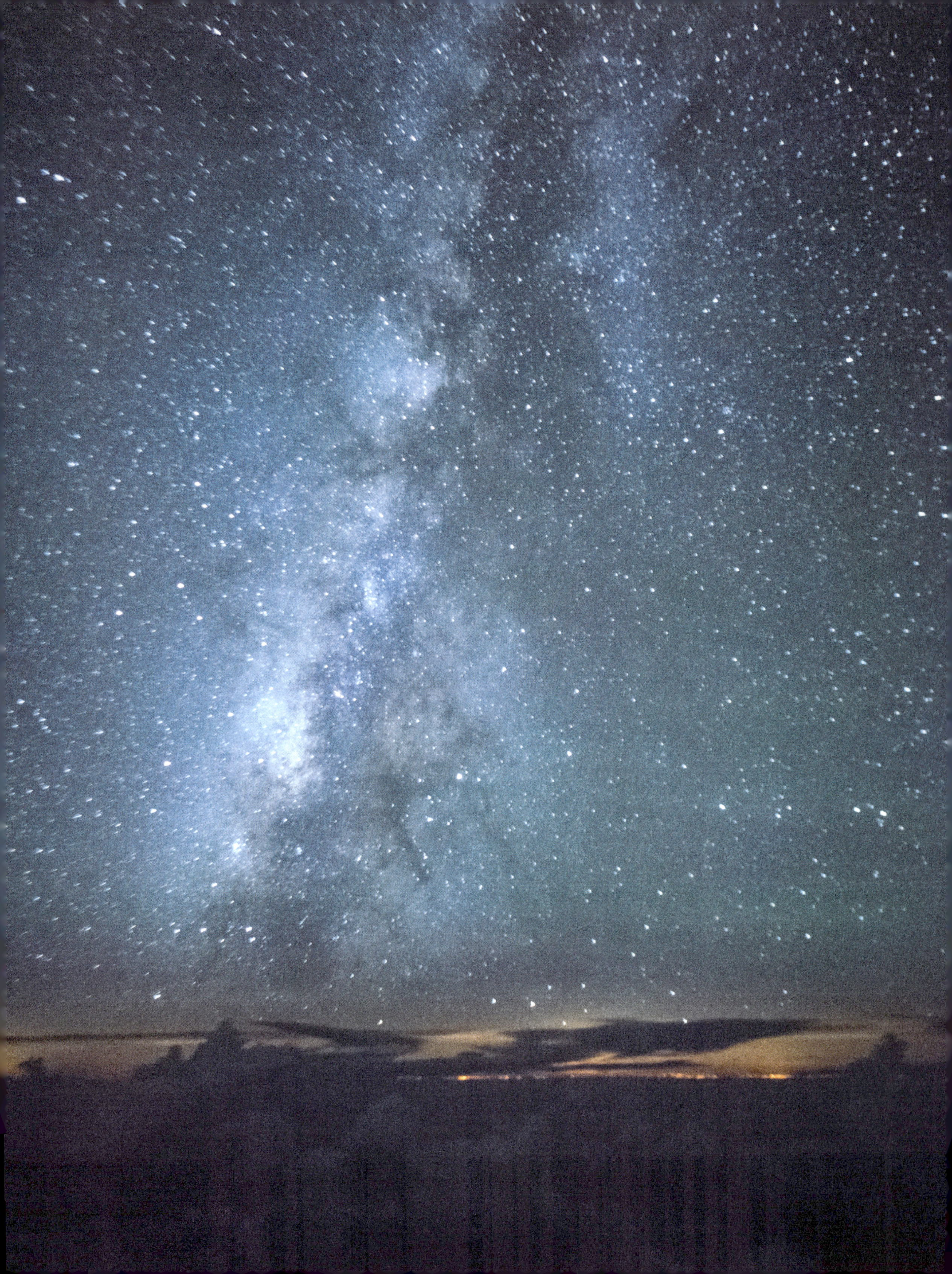

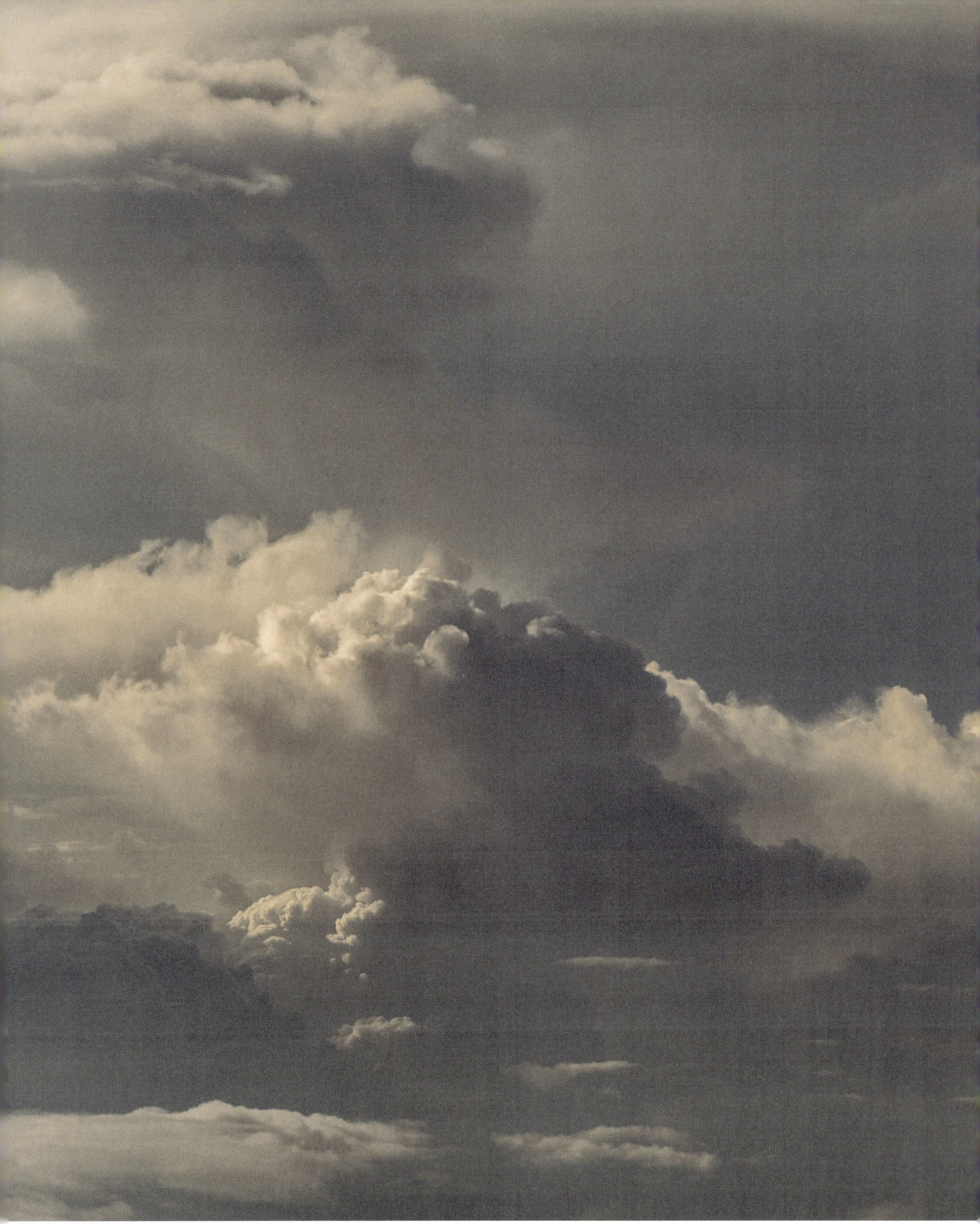

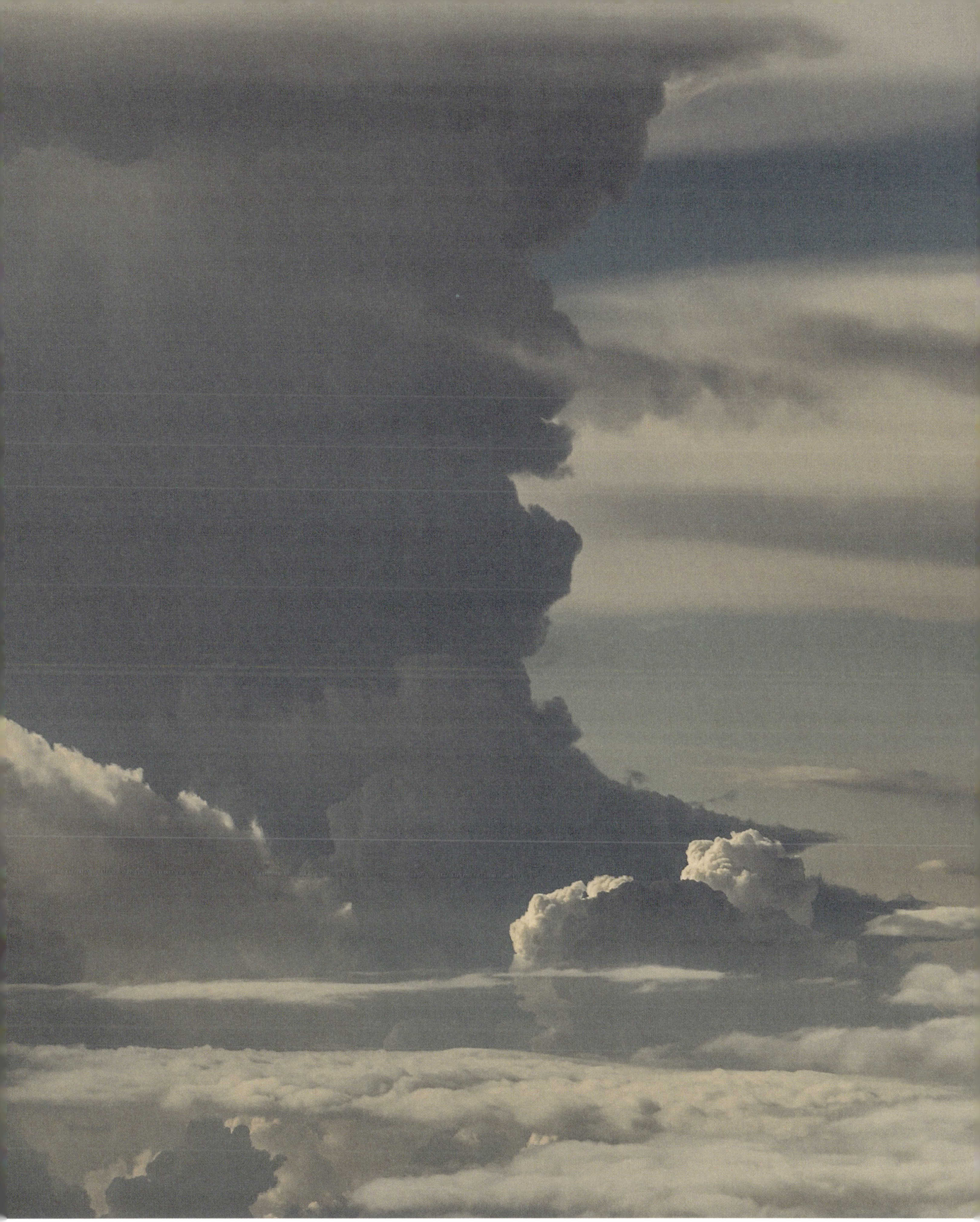

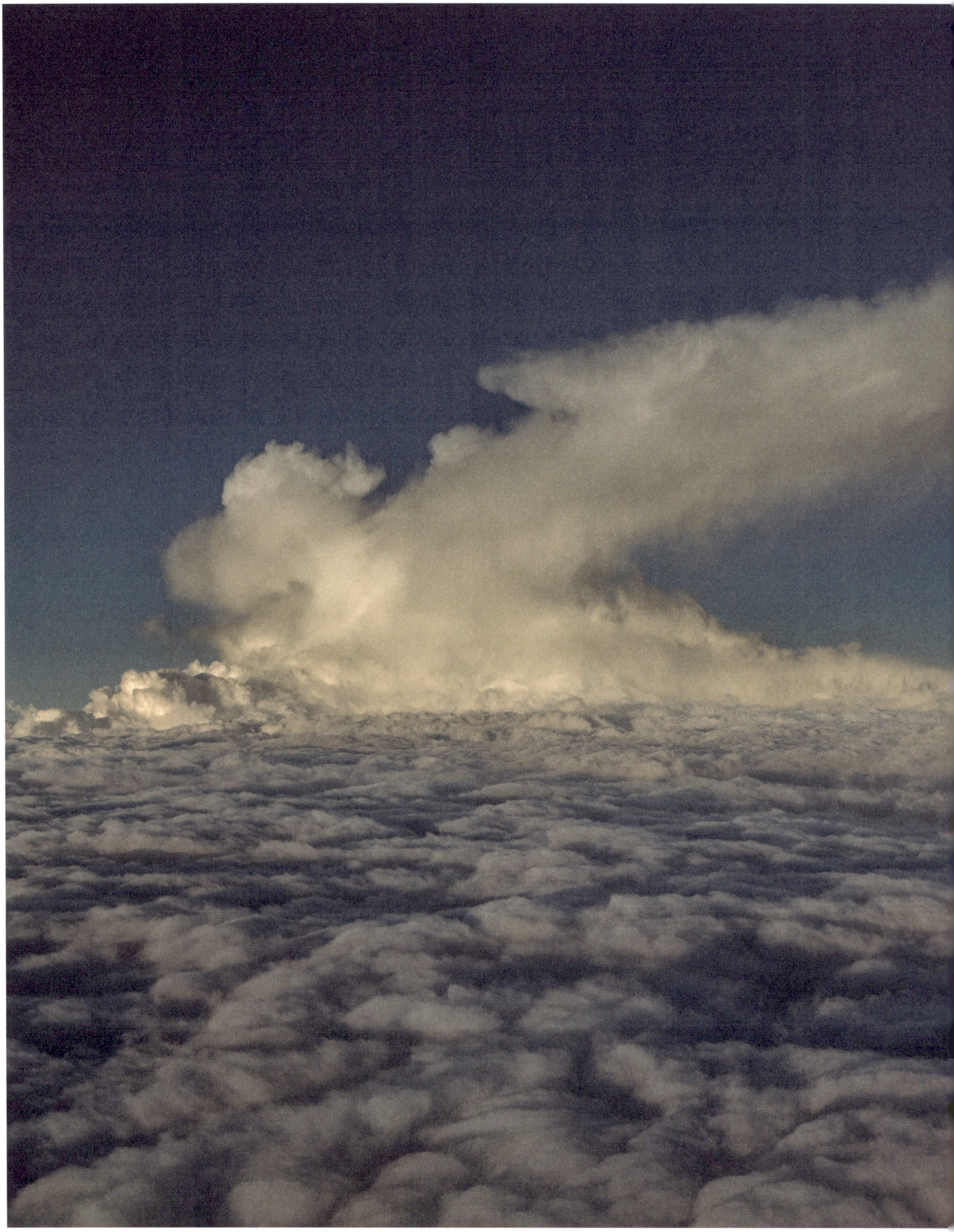

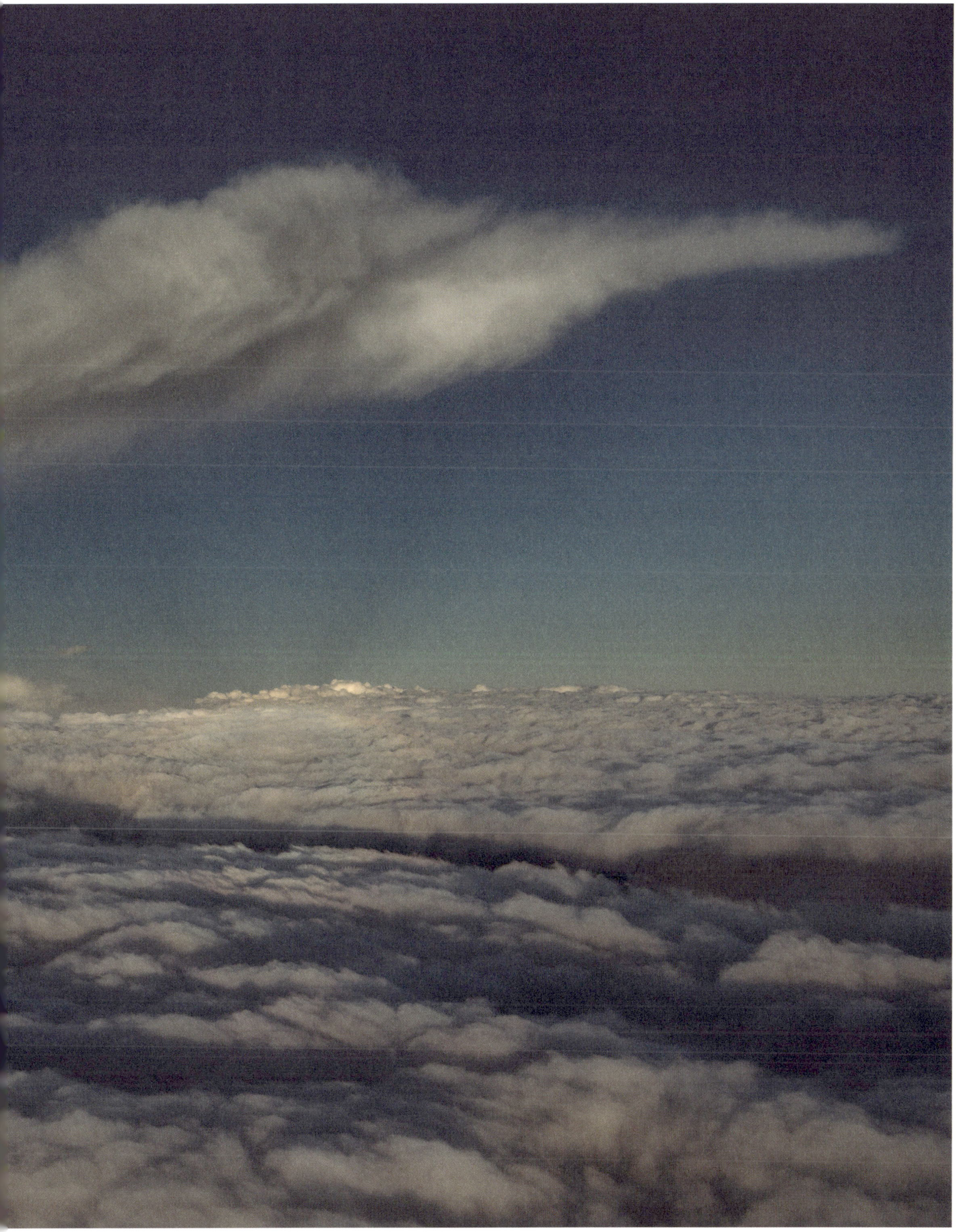

#fadetoviolet
#Horizons

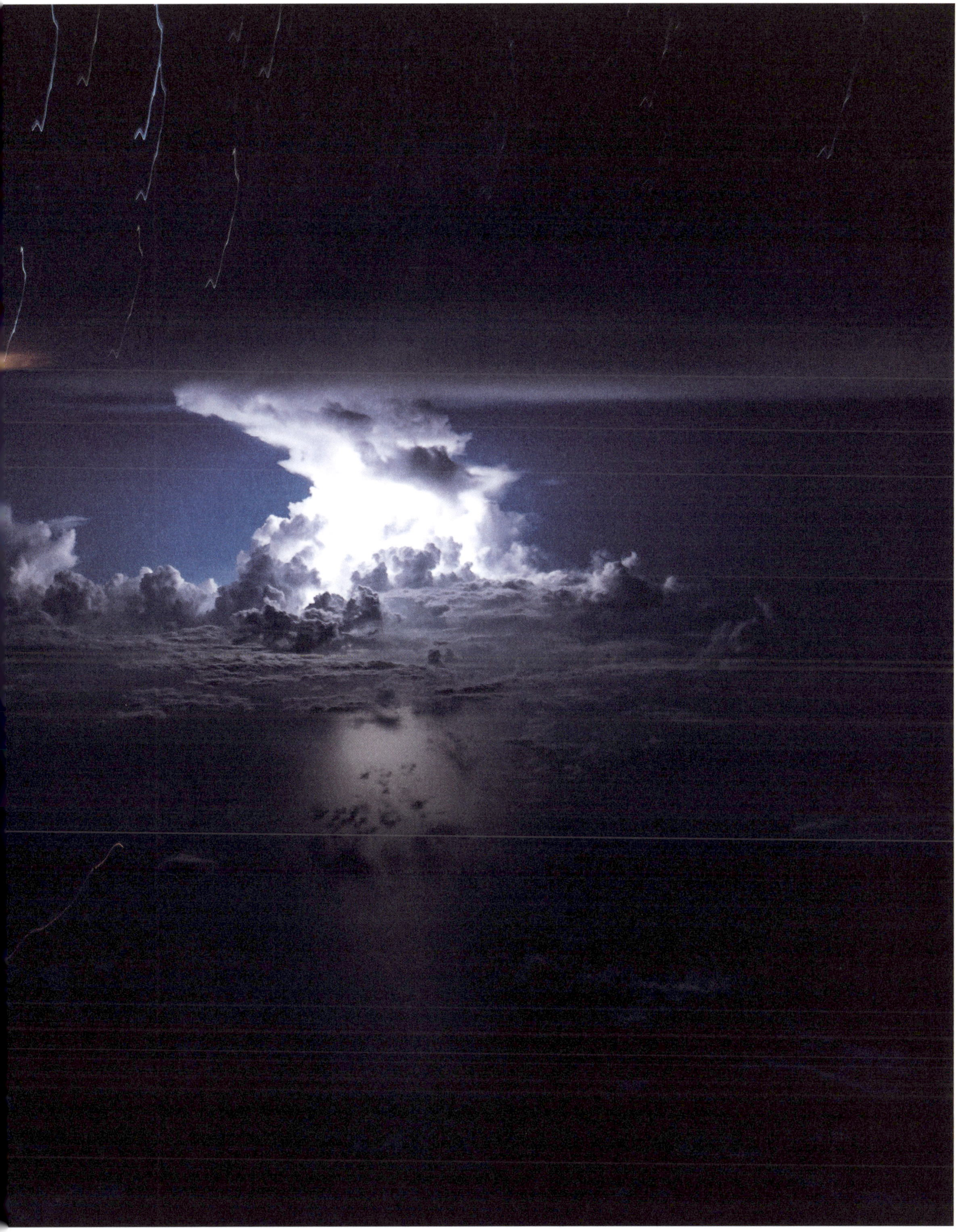

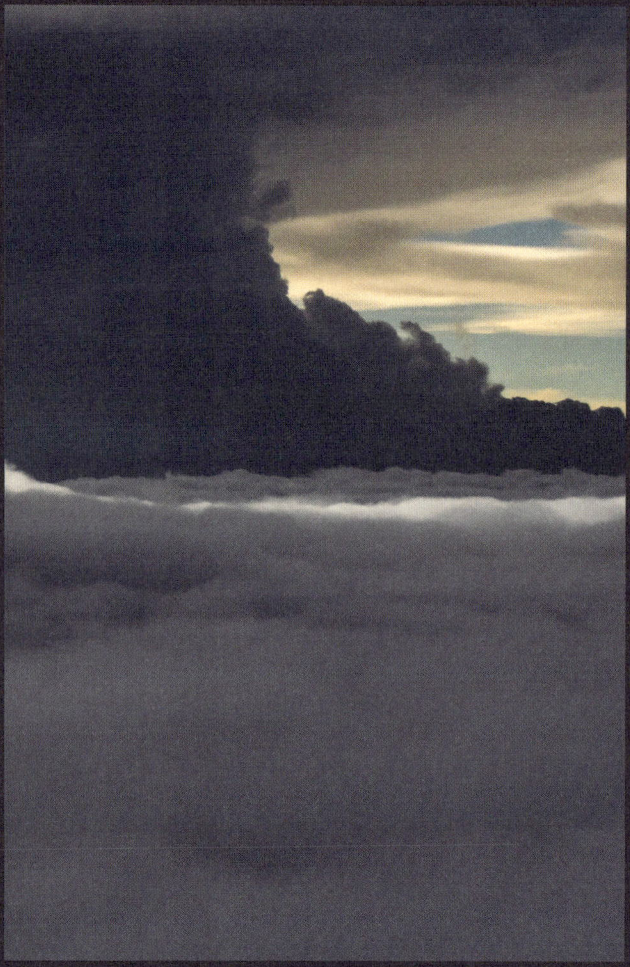

What is a storm cloud?

A storm cloud—or cumulonimbus, to give it its proper name—is a towering vertical cloud. As anyone who has ever gone mountaineering will know, air temperature drops with every mile of altitude; at 26,000 to 50,000 feet, however, where the tropopause forms the border to the stratosphere, this phenomenon is abruptly reversed and the temperature starts to rise again.

As such, this layer forms a kind of seal over storm clouds, much like the top of a pressure cooker: large amounts of wet, warm air rise upwards and start to cool as they reach the colder layers of the atmosphere; when these currents suddenly come up against air that is warmer than they are, however, they are no longer able to rise. The result is that the cold air becomes trapped under the warmth, but that it keeps on rising up from below, pushing air out sideways. This leads to the typical anvil shape of a classic cumulonimbus storm cloud—and the showers of rain or hail it can unleash hit like a hammer. Carrying up anything from 20 to 100 million tons of water, cumulonimbus are also the cloud formations associated with hurricanes: strong winds may occur both below and above them, which is why planes fly around storm clouds. With surrounding temperatures ranging from -4°F to -76°F (-20°C to -60°C), the anvil domes of these clouds consist entirely of ice crystals.

Was ist eine Gewitterwolke?

Eine Gewitterwolke (Cumulonimbus) ist eine sehr hochreichende Wolke. Mit der Höhe wird es in unserer Atmosphäre kälter. Das merkt man selbst, wenn man einen Berg besteigt und es sich mit jedem Kilometer nach oben abkühlt. Ab der Tropopause, einer Grenzschicht in acht bis 15 Kilometern Höhe, kehrt sich dieses Phänomen aber plötzlich um und es wird wieder wärmer.

Die Tropopause wirkt dadurch für die wachsenden Gewittertürme wie der Deckel eines Kochtopfs. Warme, feuchte Luft steigt in großer Menge von unten nach oben und kühlt sich dabei in kälteren Luftschichten ab. Stößt diese Luft dann plötzlich auf eine Luftschicht, die wärmer ist als sie selbst, kann sie nicht weiter aufsteigen – die kalte Luft ist gewissermaßen unter der warmen eingeklemmt. Von unten strömt aber weiterhin Luft nach, wodurch sie an der Tropopause zur Seite abgedrängt wird. Die so entstehende Wolke nimmt die Form eines Ambosses an – die klassische Gewitterwolke. Diese Wolke kann massive Schauer und Hagel mit sich bringen. Sie ist auch die Wolkenform von Wirbelstürmen und kann 20 bis 100 Millionen Tonnen Wasser enthalten. Unter- und innerhalb dieser Wolken können starke Winde auftreten, weswegen auch große Flugzeuge Gewitterwolken umfliegen. Der Amboss besteht komplett aus Eiskristallen, da es in seiner Umgebung -20 bis -60 Grad Celsius hat.

Qu'est-ce qu'un nuage orageux ?

Un nuage orageux (cumulonimbus) est un nuage à forte extension verticale. Or comme chacun peut en faire l'expérience lors d'une randonnée en montagne, l'air ambiant se refroidit à mesure que l'on s'élève. Une fois franchie la tropopause, couche d'atmosphère située entre 8 et 15 kilomètres d'altitude, ce phénomène s'inverse soudain, et il fait de nouveau plus chaud.

La tropopause joue pour les nuages orageux, en pleine extension verticale, un rôle semblable à celui d'un couvercle de casserole. Des masses importantes d'air chaud et humide s'élèvent et se refroidissent dans les couches d'air froid. Si cet air rencontre une couche d'air plus chaude que lui, il est arrêté dans sa progression, l'air froid étant pour ainsi dire piégé sous l'air chaud. Au contact de la tropopause, l'air chaud qui continue de venir du bas s'étale latéralement, et le nuage prend la forme d'enclume caractéristique des cumulonimbus. Contenant de 20 à 100 millions de tonnes d'eau, ces derniers peuvent engendrer de fortes averses et chutes de grêle. Ce sont aussi les nuages à l'œuvre dans les tornades. Comme des vents de forte intensité peuvent circuler dans et entre ces nuages, même les gros avions les contournent. L'enclume est entièrement constituée de cristaux de glace car la température de l'air qui l'entoure se situe entre -20 et -60° Celsius.

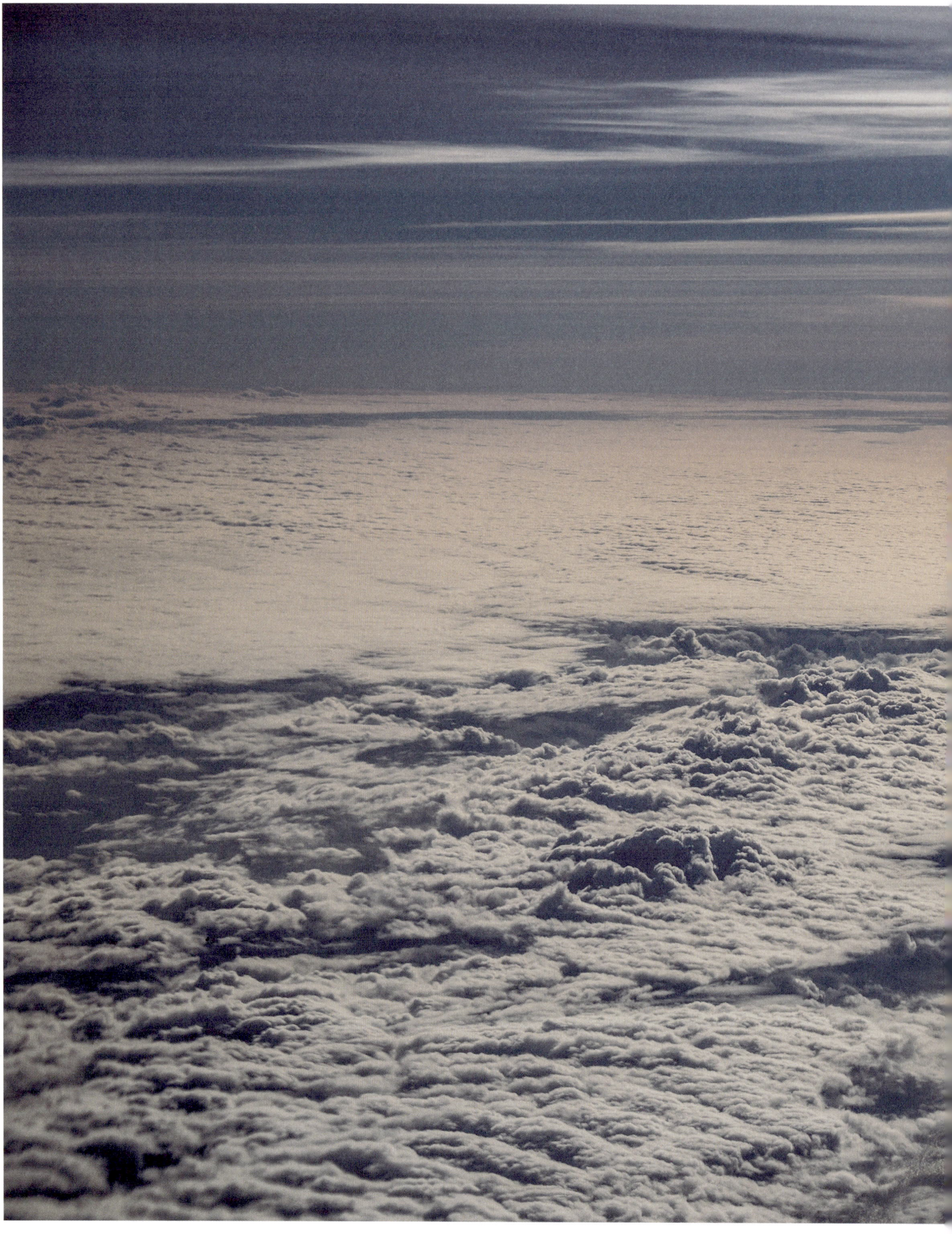

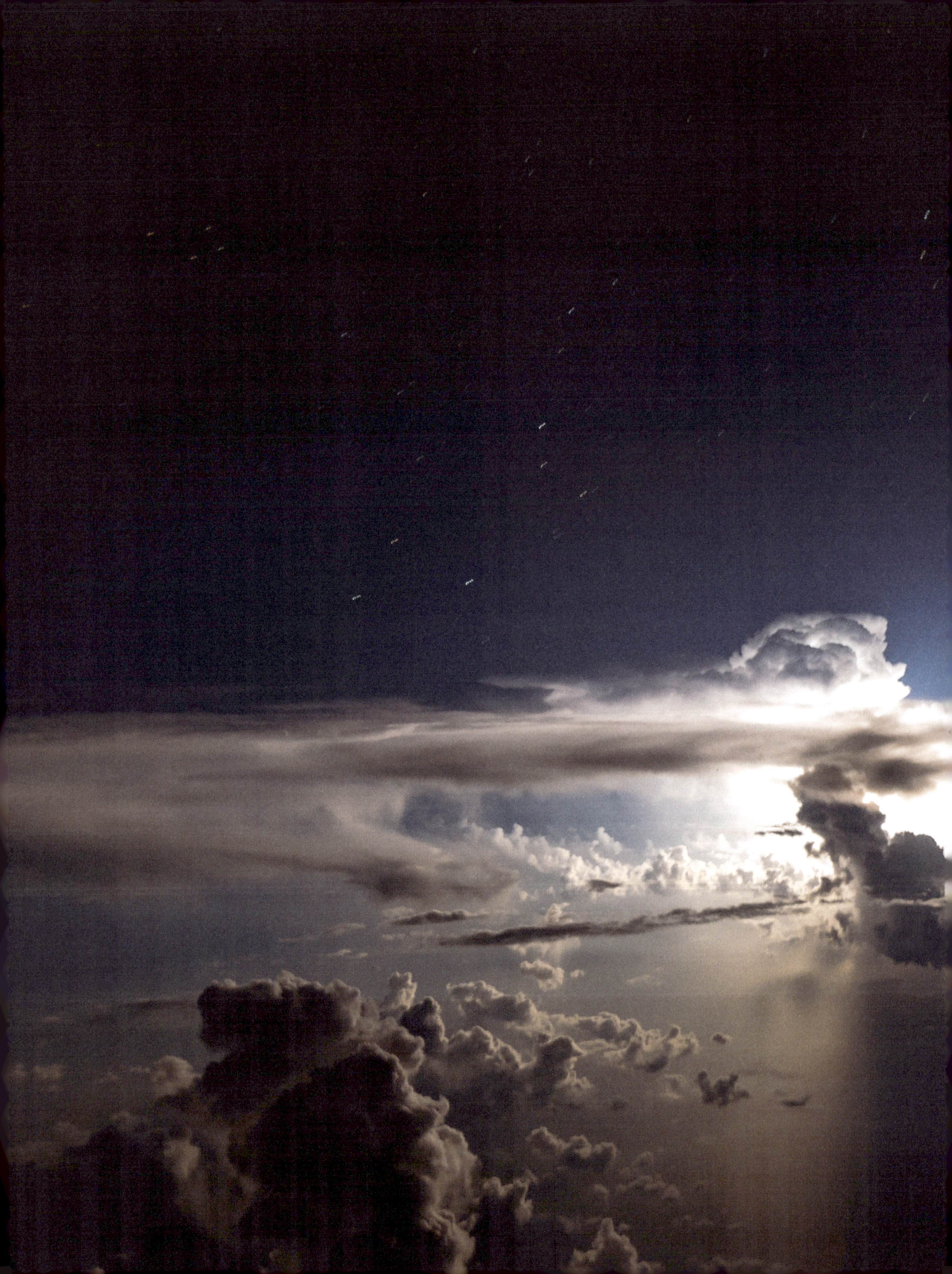

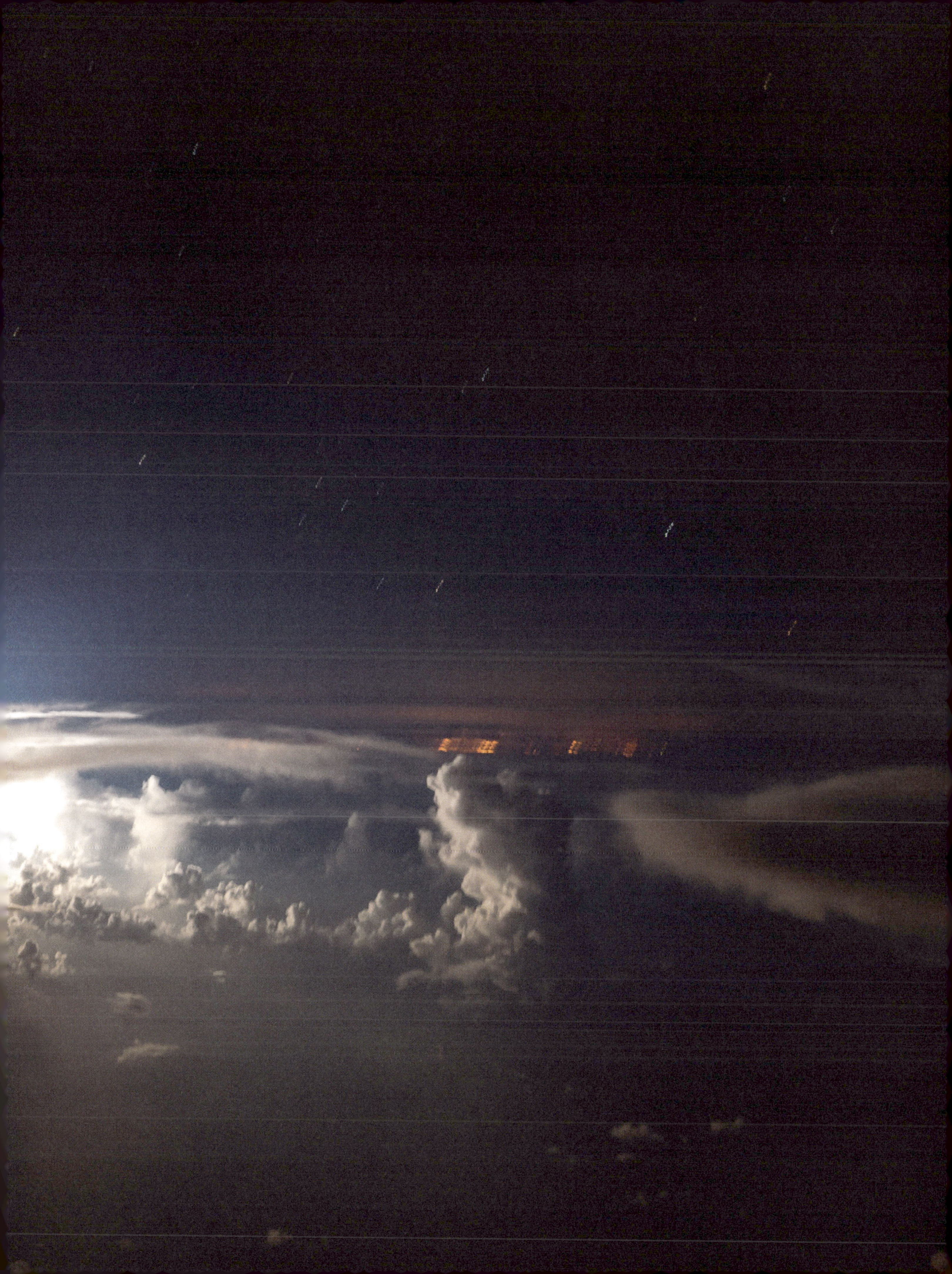

What kind of storms are there?

As meteorologists, we differentiate between two main types of storms: air-mass thunderstorms and multicell or cluster storms. The latter kind occurs when two areas of air at different temperatures meet—this is called a weather front. Storms tend to develop along cold fronts, which contain more wind and more energy than warm fronts; the colder air pushes its way under the warmer mass, forcing it to rise—and creating storm clouds. This kind of cluster storm can strike in several places at once and tends to hit large cities especially hard: urban areas are heat islands and maximise the power of storms passing over.

Air-mass thunderstorms, meanwhile, include the typical heat storms of the summer months; they generally do not last longer than an hour and happen when, over the course of an afternoon, cumuliform clouds ("cauliflowers") develop into large storm piles—which then tend to discharge rain, thunder and lightning in a very limited area. Just a few hundred yards away, the weather can remain summery with some light cloud cover while, under the storm, heavy showers and even hailstorms hammer down. Meteorologists have a particularly tough time

Welche Arten von Gewittern gibt es?

Wir Meteorologen unterscheiden im Prinzip zwei Arten von Gewittern: Luftmassen- und Frontgewitter. Erstere entstehen an der Grenze zweier unterschiedlich warmer Luftmassen, an einer Front. Meist bilden sich Gewitter entlang von Kaltfronten, hier sind mehr Wind und mehr Energie in der Atmosphäre als bei ihren warmen Pendants. Die kühle Luft der Kaltfront schiebt sich dabei unter die wärmere Luft, wodurch diese aufsteigt – es entstehen Gewitterwolken. Bei Frontgewittern kann es viele Orte zur gleichen Zeit treffen, wobei es sich über Großstädten häufig besonders heftig entlädt. Städte sind Wärmeinseln und verleihen den Gewittern so noch zusätzliche Power.

Zu den Luftmassengewittern zählt man die im Sommer der gemäßigten Breiten typischen Wärmegewitter. Ihre Lebenszeit erreicht meist nur 60 Minuten. Im Laufe des Nachmittags können sich aus Blumenkohlwolken große Gewitterberge entwickeln. Diese regnen, blitzen und donnern sich dann meist nur über einem geografisch relativ kleinen Gebiet aus. Wenige hundert Meter weiter schon bleibt es häufig bei nur leicht bewölktem Sommerwetter. Dennoch kann es an Ort und Stelle schon ordentlich rumsen, platzregnen

Quels sont les différents types d'orage ?

Les météorologues distinguent en gros deux types d'orages : les orages orographiques (ou convectifs) et les orages frontaux (ou cycloniques). Ces derniers se forment à la frontière de deux masses d'air de températures différentes, sur un front. Et ce, de préférence sur un front froid, où il y davantage de vent et d'énergie dans l'atmosphère que sur un front chaud. L'air frais du front froid s'infiltre sous l'air plus chaud, qui s'élève en entraînant la formation de nuages orageux. Les orages frontaux peuvent toucher de nombreux endroits simultanément, se déchargeant souvent de manière particulièrement violente sur les grandes villes. Les villes sont en effet des îlots de chaleur qui accroissent l'intensité des orages.

Sous-catégorie des orages orographiques (ou convectifs), les orages thermiques ou de chaleur sévissent surtout en été. Ils ne durent en général pas plus de 60 minutes. Au cours de l'après-midi, des nuages en forme de chou-fleur (cumulonimbus calvus) peuvent devenir de gros nuages orageux, qui engendrent pluie, éclairs et tonnerre en général au-dessus d'une zone géographique relativement limitée. Il vente, il pleut et il y grêle tant et plus. Quelques centaines de mètres plus loin, le

predicting this kind of highly localised storm activity. The only way to alert those affected is to use now-casting, which means evaluating live radar images and dispatching specific warnings just before the storm develops; the maximum horizon for this type of forecast is 30 minutes.

In some specific volatile conditions, storms can start to interact and mutually reinforce one another, combining into long-lasting, deeply powerful weather events. This kind of mega-storm can take on gigantic proportions, reaching 33,000 to 50,000 feet into the sky; the interiors of these storm clouds house incredible amounts of energy which is generally discharged in a highly damaging fashion—lightning-intensive thunderstorms, tornadoes, hailstones the size of tennis balls, or torrential rain causing flooding are often the result. Experts classify these types of events into multicell, supercell, and mesoscale convective systems: what they all have in common is that, if one is approaching, you're best off staying indoors.

und hageln. Für Meteorologen ist es besonders schwer zu sagen, wo genau diese Wärmegewitter auftreten werden. Im Prinzip lassen sich solche Vorhersagen nur durch „Now-Casting" vorhersagen – dabei begutachtet man Radarbilder und kann maximal 30 Minuten vor dem Gewitter eine konkrete Warnung herausgeben.

Unter bestimmten dynamischen Bedingungen können sich Gewitterherde verstärken, wodurch sie zu langlebigen, heftigen Unwettern (Schwergewitter) werden. Deren Ausdehnung kann gigantisch sein: zehn bis 15 Kilometer ragen diese Gewitterformationen in den Himmel. Im Inneren dieser Wolken herrscht unglaubliche Energie, die sich leider meist auf tragische Weise entlädt: Sehr blitzintensive Gewitter, Tornados, tennisballgroße Hagelkörner oder heftigste Regenmengen in sehr kurzer Zeit, die zu Überschwemmungen führen, sind meist die Folge. Die Experten unterscheiden die langlebigen Gewitterherde noch nach Multizellen, Superzellen und mesoskaligen Konvektionskomplexen: Wetterphänomene, bei denen man das Haus besser nicht verlassen sollte.

temps reste estival, et le ciel juste légèrement couvert. Les météorologues peuvent difficilement prévoir l'endroit précis où ces orages de chaleur vont éclater. En principe, ces orages ne peuvent être détectés que grâce à l'observation d'images radar en temps réel, ce qui ne permet pas de lancer une alerte plus de 30 minutes avant qu'ils n'éclatent.

Dans certaines conditions dynamiques, des foyers orageux sont susceptibles de se développer et de devenir de violentes intempéries (orages) dont l'extension verticale peut atteindre la hauteur considérable de 10 à 15 kilomètres. Ces nuages recèlent une énergie incroyable qui se décharge hélas en général de manière dramatique : il s'ensuit en général des éclairs, des tornades, des grêlons gros comme des balles de tennis ou des pluies torrentielles en très peu de temps responsables d'inondations. Les spécialistes classent les foyers orageux persistants en orages multicellulaires, supercellulaires et en complexes convectifs de méso-échelle. Autant de phénomènes météorologiques qui invitent à rester chez soi.

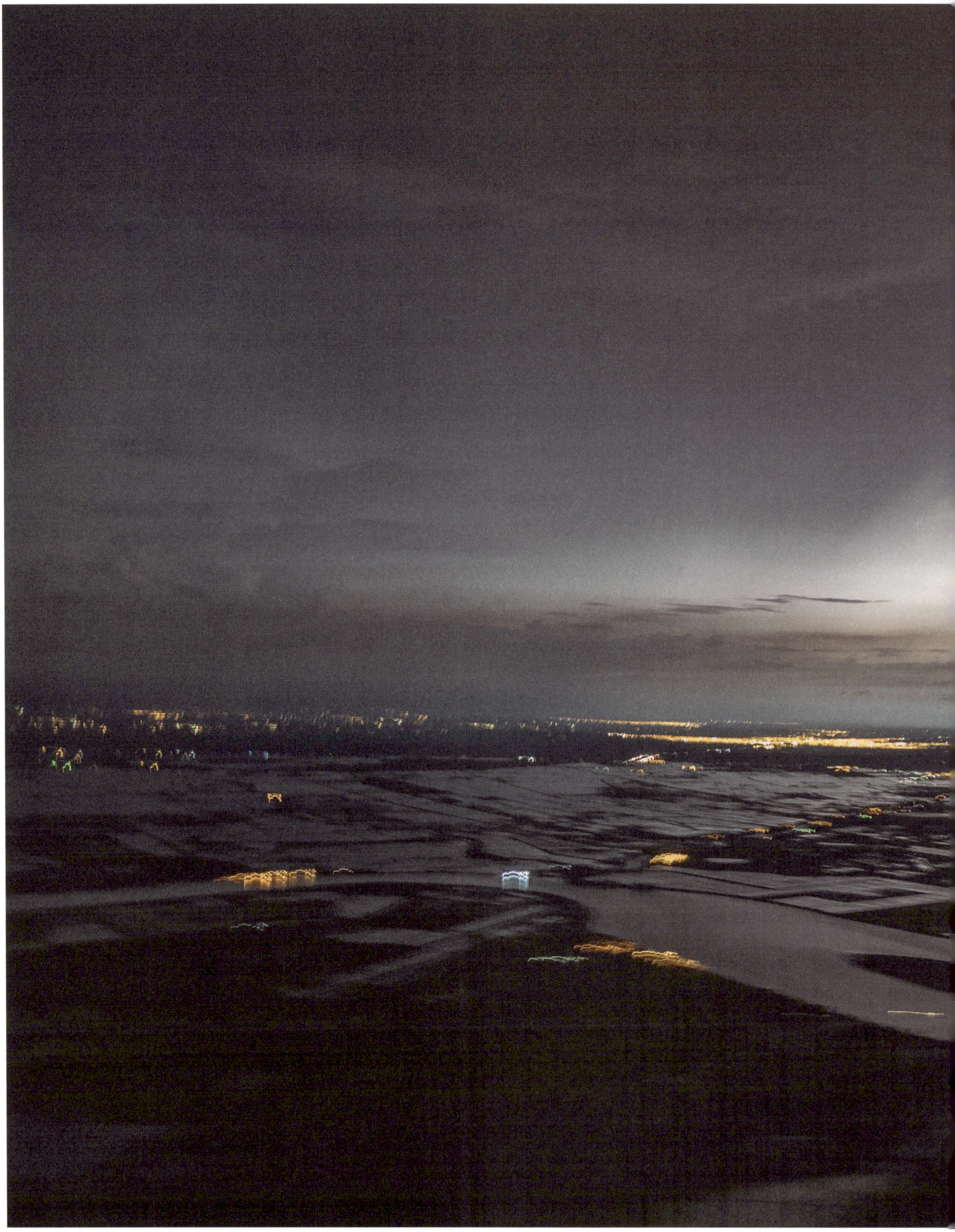

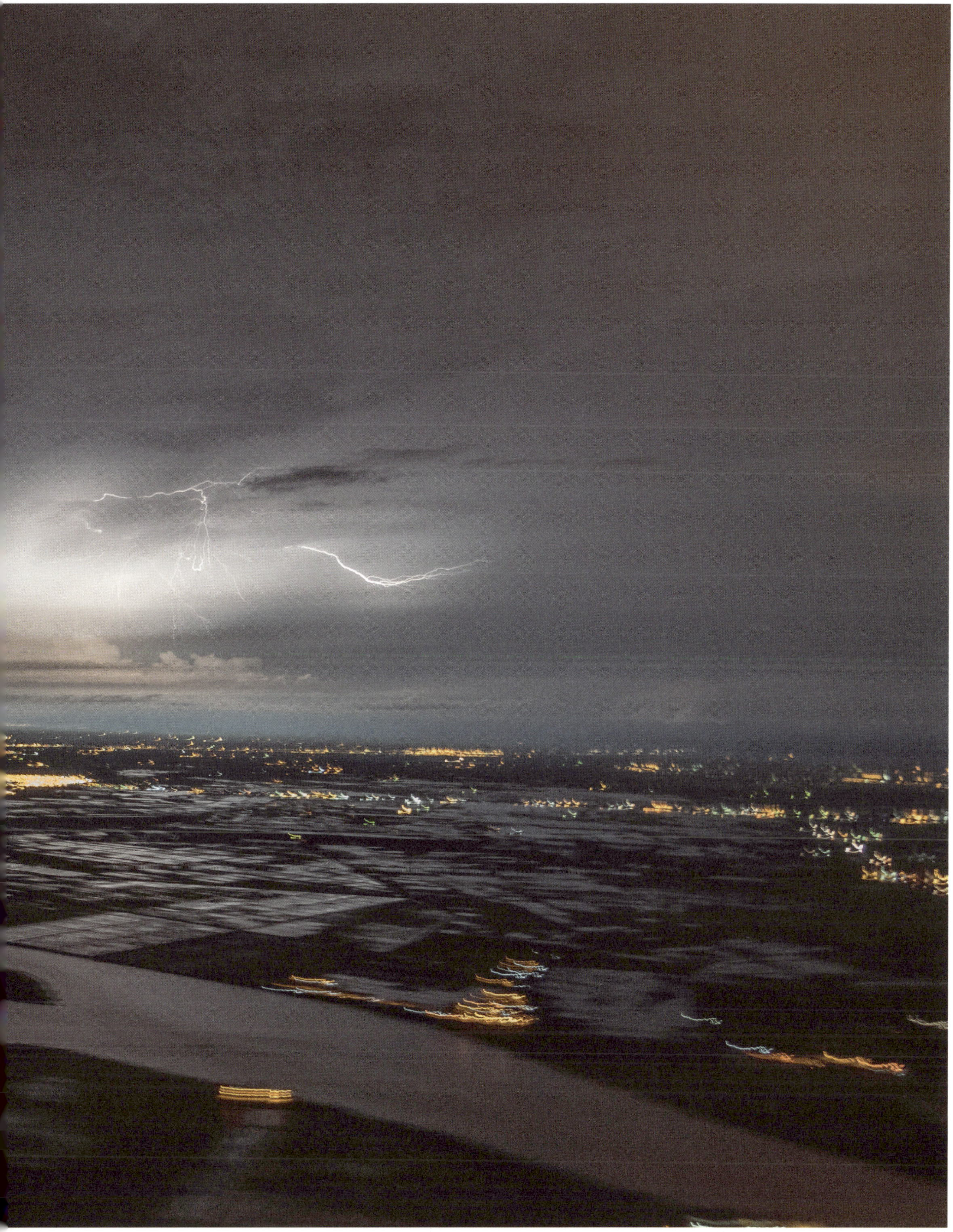

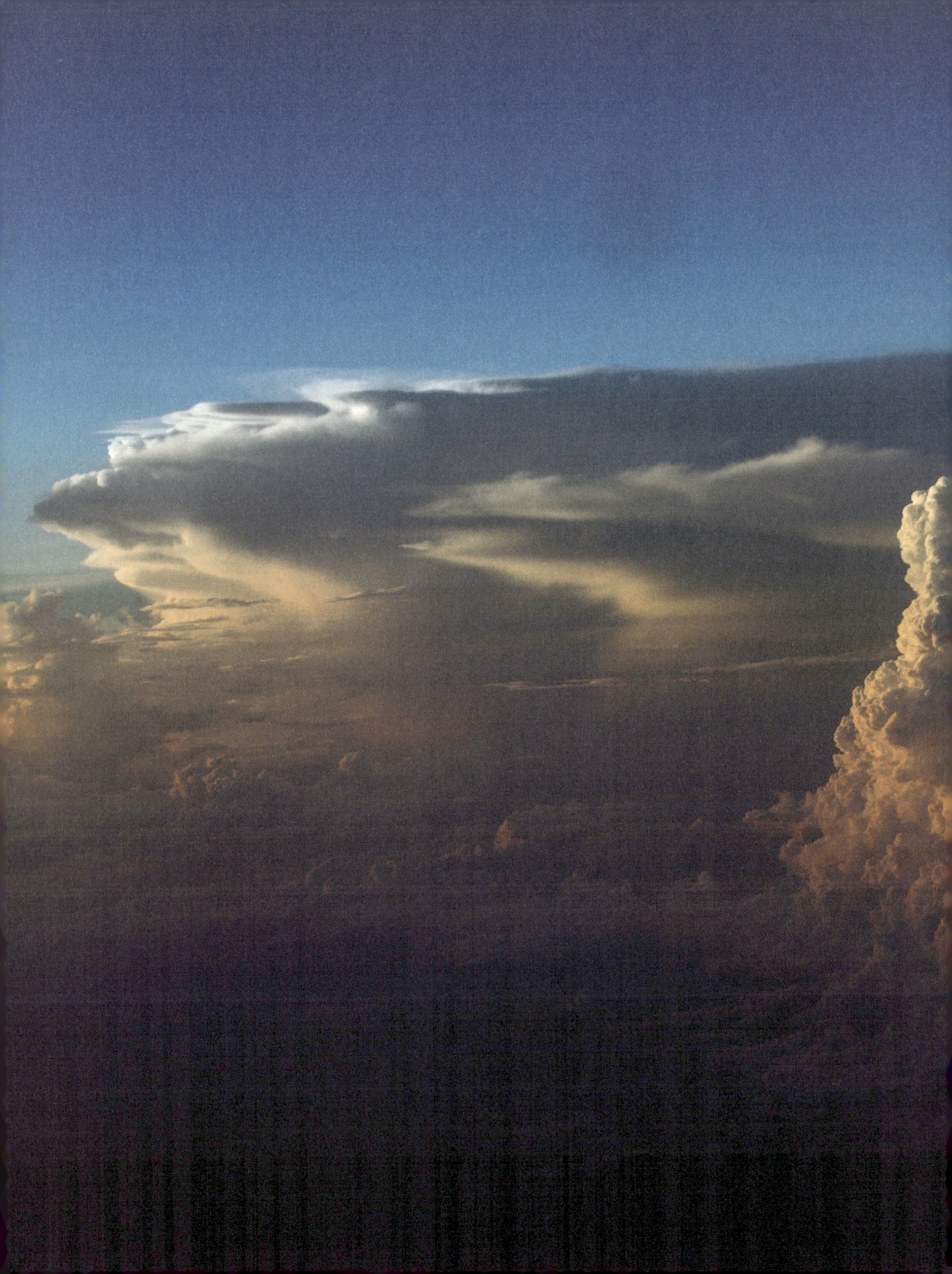

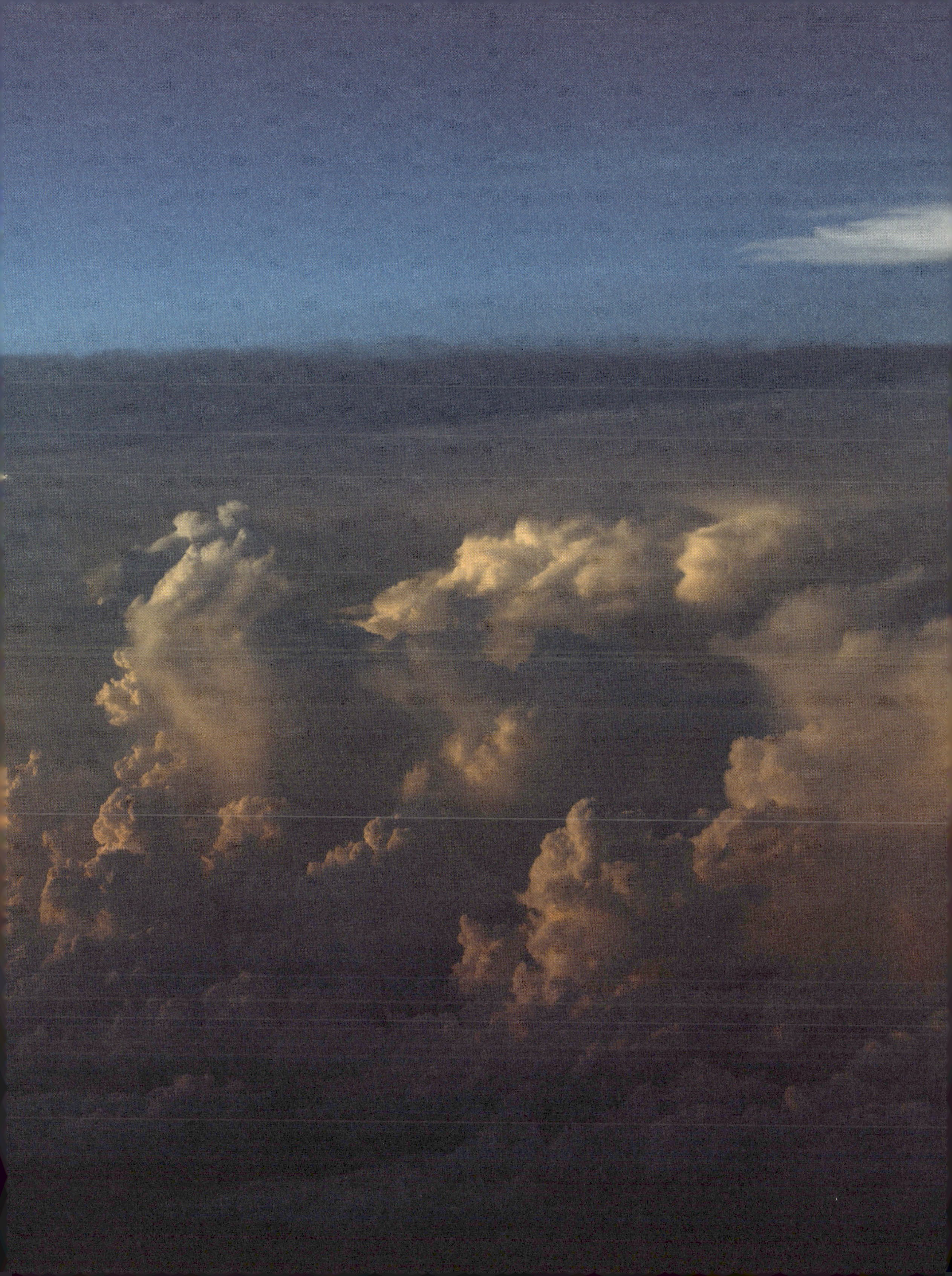

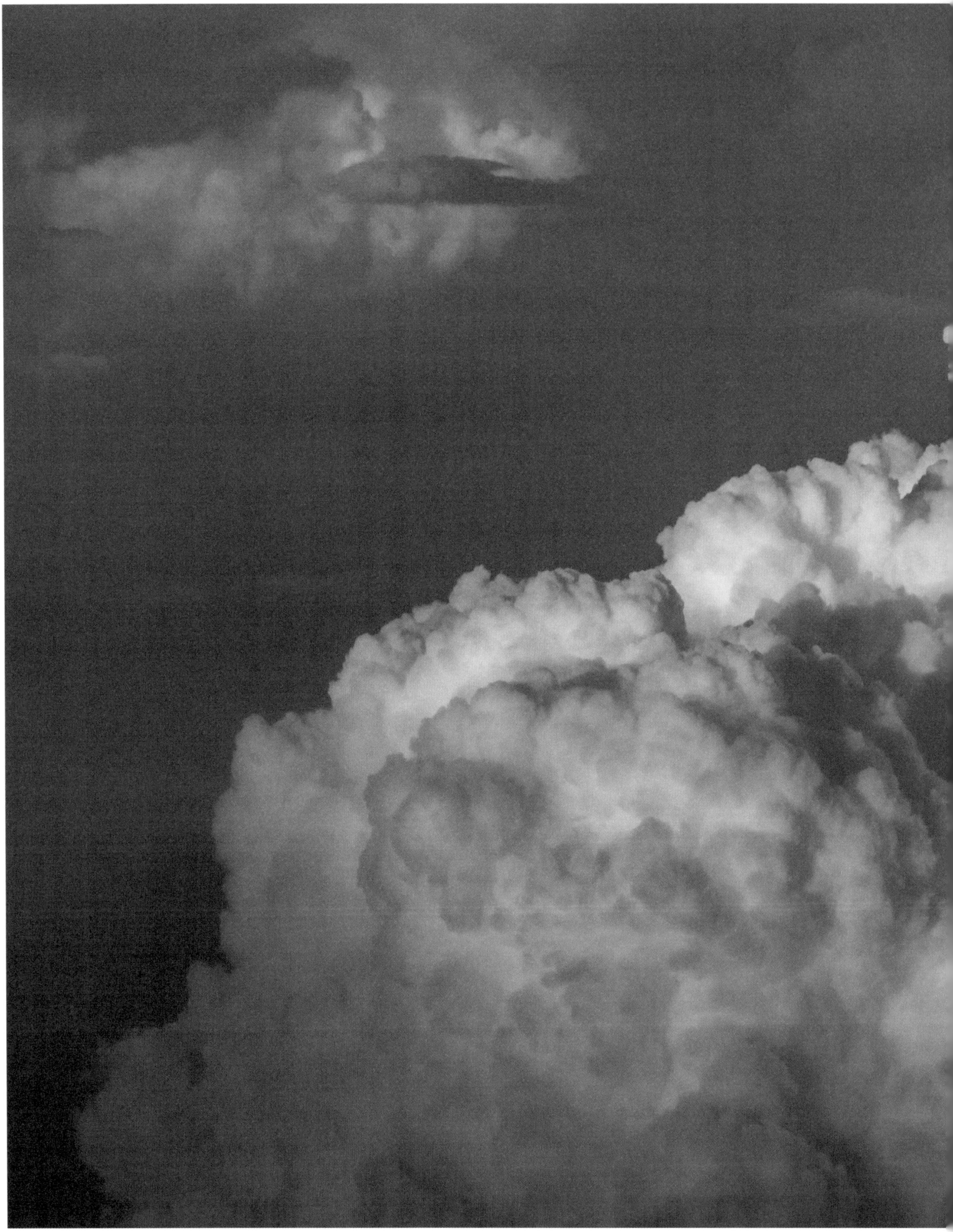

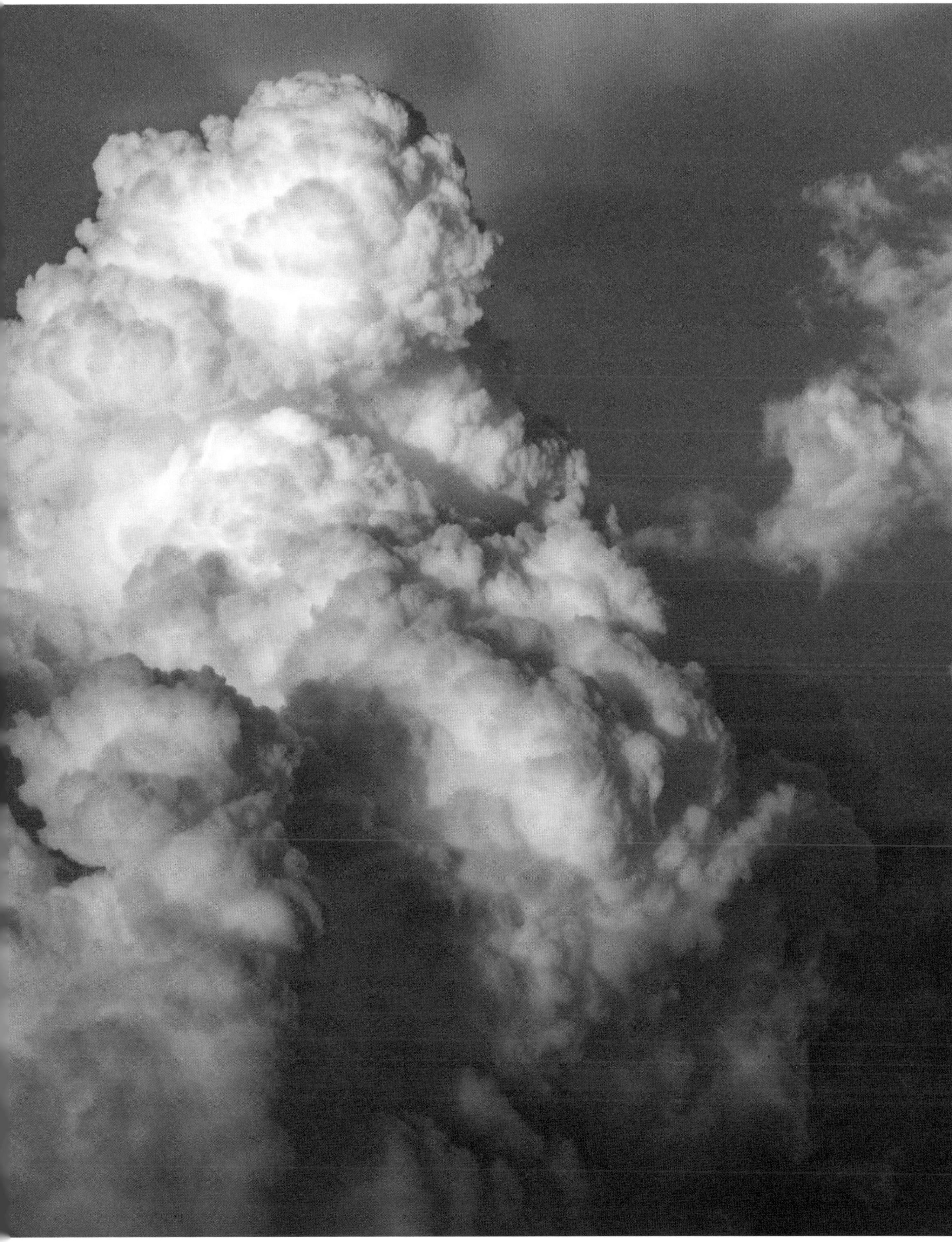

Why are storms in the mountains so dangerous?

For the people caught up in them, storms can be even more dangerous in the mountains than elsewhere: the speed at which the weather turns at higher altitudes means that there is often precious little time to seek shelter, so checking the latest weather forecasts before heading out is an absolute must. Once out, mountaineers should keep a close eye on conditions. In summer, it's advisable to start early in order to avoid the storms which can develop during the afternoon heat.

Lightning sighted without audible thunder means that the storm is at least 11 miles away (the maximum distance its soundwaves can cover). From this point on, it is crucial to track the direction in which the storm clouds are moving: if the space of time between seeing lightning and hearing thunder gets shorter, that means the storm is approaching; if the pause between the two is less than three seconds, that's a sign of imminent danger—lightning could strike in the immediate vicinity at any moment. Anyone who hasn't made it to a shelter with a lightning rod by this point should go to ground, preferably in a hollow and curl up into a ball. Until the storm has past, there is both, a risk of being

Warum sind Gewitter in den Bergen so gefährlich?

Gewitter in den Bergen können für den Menschen noch gefährlicher werden als im Flachland. Denn hier schlägt das Wetter besonders schnell um, häufig bleibt kaum Zeit, sich in Sicherheit zu bringen. Bevor man also eine Bergtour wagt, sollte unbedingt der aktuelle Wetterbericht eingeholt werden. Auch ist es ratsam, während der Wanderung regelmäßig den Himmel zu beobachten. Vor allem im Sommer ist ein früher Start sinnvoll, denn in den Nachmittagsstunden bilden sich häufig Wärmegewitter.

Wenn man einen Blitz sieht, aber noch keinen Donner hört, ist das Gewitter noch mindestens 18 Kilometer weg, denn ab dieser Entfernung wird der Schall verschluckt. Ganz wichtig ist es jetzt, die Zugbahn der Gewitterwolken im Auge zu behalten. Kommt das Gewitter näher, werden die zeitlichen Abstände zwischen Blitz und Donner kleiner. Brandgefährlich wird es bei einem Abstand von weniger als drei Sekunden – dann ist mit einem Einschlag in der Nähe zu rechnen. Schafft man es in so einem Fall nicht mehr in eine Hütte mit Blitzableiter oder in eine größere Höhle, sollte man sich in einer Kuhle möglichst klein zusammenkauern. Die Gefahr besteht jetzt nicht

Pourquoi les orages sont-ils aussi dangereux en montagne ?

En montagne, les orages peuvent se révéler encore plus dangereux pour l'Homme qu'en plaine. Les conditions météo peuvent y changer si brusquement que l'on n'a souvent guère le temps de se mettre à l'abri. Avant d'entreprendre une randonnée en montagne, il faut absolument consulter les prévisions météo en temps réel. Il est aussi indiqué d'observer le ciel régulièrement pendant la randonnée et de partir au petit matin, surtout en été, saison où des orages de chaleur tendent à se former dans le courant de l'après-midi.

Quand on voit un éclair mais qu'on n'entend pas encore de tonnerre, l'orage est éloigné d'au moins 18 kilomètres, distance à laquelle le son est absorbé. À partir de ce moment-là, il est capital de surveiller la trajectoire des nuages orageux. Si l'orage approche, l'intervalle de temps séparant l'éclair du tonnerre se réduit. Quand cet intervalle est inférieur à trois secondes, le danger est imminent : il faut s'attendre à un impact de foudre à proximité. Si l'on n'a plus le temps de s'abriter dans un refuge doté d'un paratonnerre ou dans une grotte suffisamment grande, il faut s'accroupir dans un creux en se faisant

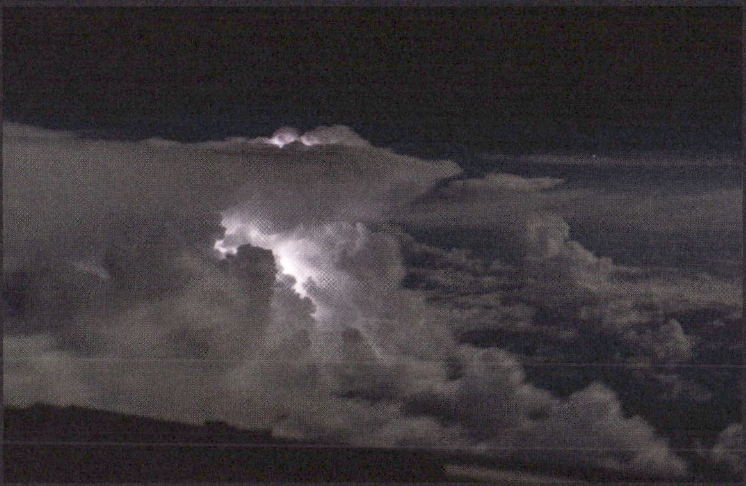

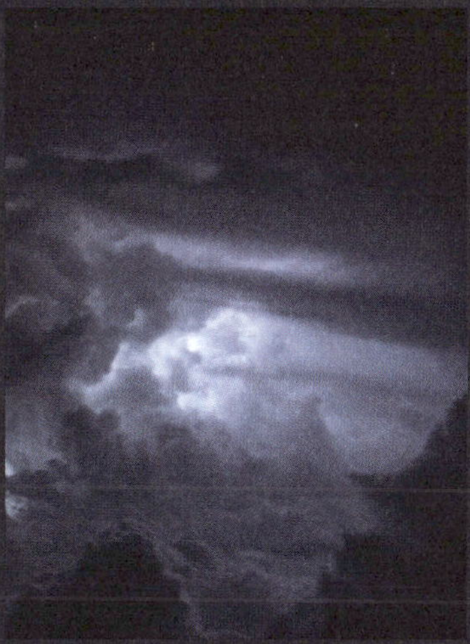

struck directly by lightning and of falling victim to creeping current: this occurs when a bolt of lightning hits the ground nearby and the current spreads outwards. Minimising direct contact with the ground is advisable here, for instance by huddling on rucksacks.

Any metallic objects should be cast away, but—contrary to rumour—mobile phones don't increase the risk of being struck by lightning and so can be kept stowed away.

Storms are frequently accompanied by downpours which can, in turn, lead to falling rocks or swell mountain creeks to break their banks; high winds may break branches off trees, too, so it's best to avoid seeking shelter in the woods.

nur darin, direkt vom Blitz getroffen zu werden. Gefährlich sind auch die sogenannten Kriechströme. Dabei schlägt der Blitz in den Boden ein, wodurch sich der Strom kreisförmig ausbreitet. Am besten setzt man sich daher auf den Rucksack, um die eigene Kontaktfläche mit dem Boden möglichst klein zu halten.

Metallische Gegenstände sollten weggeworfen werden, das Handy kann man im Rucksack verstauen. Entgegen anderslautender Gerüchte erhöht ein Handy nicht die Wahrscheinlichkeit, vom Blitz getroffen zu werden.

Mit Gewittern kommt häufig heftiger Regen, er kann zu Steinschlägen führen, Flussläufe steigen rasend schnell über die Ufer. Sturmböen lassen Äste abbrechen – im Wald Schutz zu suchen, ist also keine gute Idee.

le plus petit possible. Au danger d'être frappé directement par la foudre s'ajoute celui des courants de fuite, qui se propagent concentriquement autour de l'impact de foudre. Le mieux est de s'asseoir sur son sac à dos pour limiter la surface de contact entre son propre corps et le sol.

S'il faut jeter loin de soi les objets métalliques, il suffit de ranger son portable dans son sac à dos car, contrairement à une croyance répandue, cet appareil n'accroît pas le risque d'être frappé par la foudre.

Les orages s'accompagnent souvent de pluies diluviennes, qui peuvent entraîner des éboulements et des crues subites des cours d'eau. Les rafales de vent pouvant casser des branches, il n'est pas judicieux de chercher refuge dans les bois.

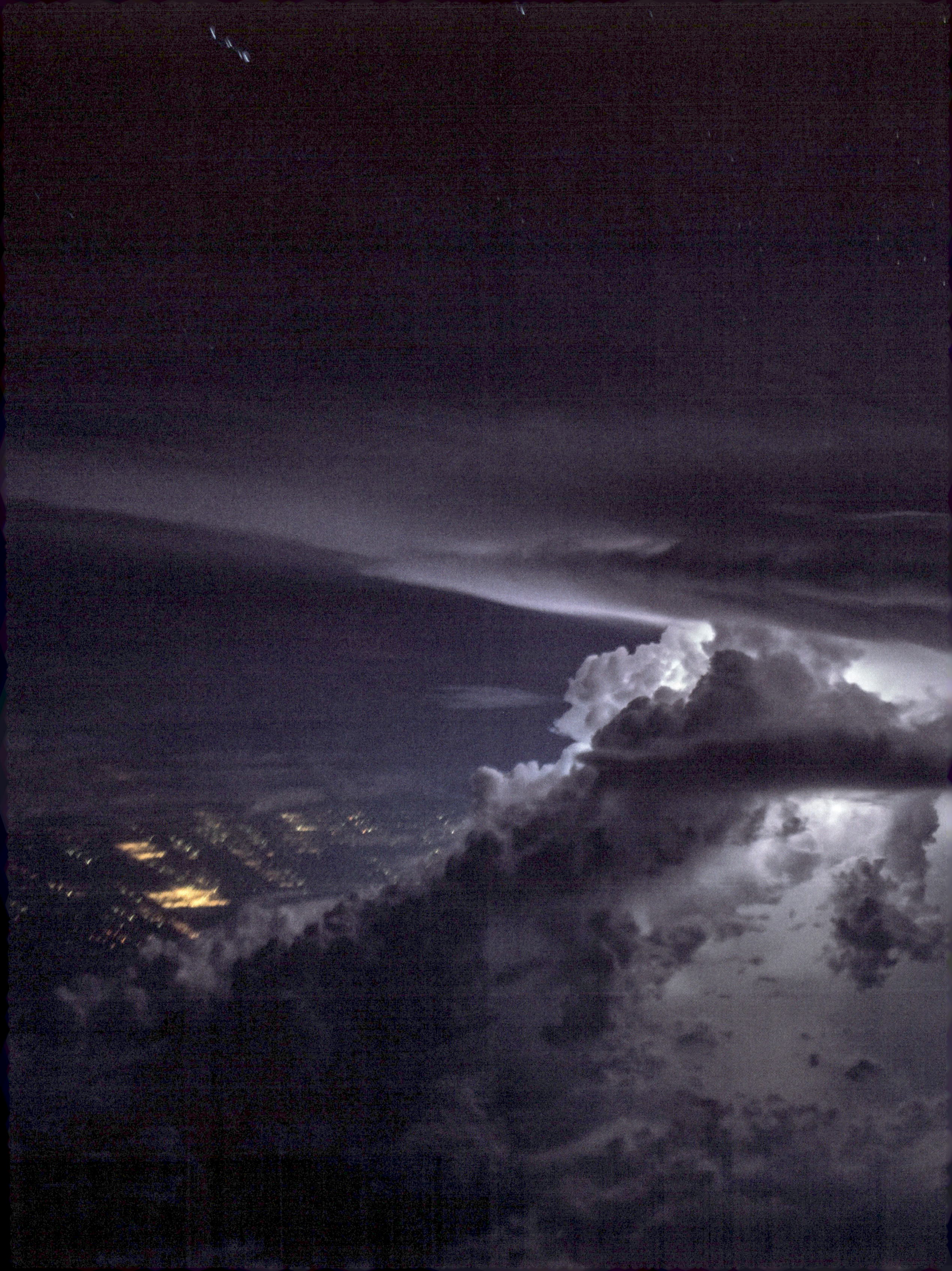

What is wind?

The Oxford English Dictionary defines wind as "air in motion, a state of movement in the air", that is indeed the core of what wind is. Mostly, however, people talk about winds when currents of air are noticeably moving in one direction and are stronger than the movements of the air we are subject to every—at least, when we are outside.

HOW DOES WIND HAPPEN, THOUGH?

As with so many other meteorological phenomena, it is the sun which is pulling the strings here. At the equator, it warms the earth to a similar degree all year round, blazing high above the horizon and rarely letting temperatures drop far below 85°F (30°C); at the poles, however, the sun doesn't shine at all for half a year at a time and, even when it is shining there, temperatures rarely make it far beyond freezing.

Nature, however, likes to try and even things out, moving the cooler air from the poles towards the equator and bringing warmer air back the other way. That is the big picture in broad terms.

On a smaller scale, meanwhile, landmasses and bodies of water heat up at different speeds and to different temperatures under the influence of the sun.

Was ist Wind?

Definiert ist Wind als eine „gerichtete, starke Luftbewegung in der Erdatmosphäre". Einfach ausgedrückt: Wind ist Luft, die sich in eine bestimmte Richtung bewegt. Dabei sind wir – zumindest außerhalb von Häusern – tagtäglich von Wind umgeben, machen uns darüber meist aber kaum Gedanken.

ABER WIE ENTSTEHT WIND EIGENTLICH?

Auch hier hat die Regie mal wieder die Sonne. Sie erwärmt die Erde unterschiedlich stark: Am Äquator ist sie das ganze Jahr über fleißig und steht hoch am Horizont, hier fällt das Thermometer kaum einmal gravierend unter 30 Grad Celsius. An den Polen dagegen scheint die Sonne ein halbes Jahr lang gar nicht, und das andere halbe Jahr ist es mit Temperaturen um den Gefrierpunkt auch nicht gerade kuschelig.

Da die Natur aber versucht, die Dinge im Gleichgewicht zu halten, wird durch Wind die kalte Luft von den Polen Richtung Äquator befördert, während die warme Luft den umgekehrten Weg geht.

Das geschieht im Großen. Im Kleinen erwärmen sich die Landmasse und das Meer durch Sonneneinstrahlung unterschiedlich schnell und stark. Luft über

Qu'est-ce que le vent ?

En météorologie, le vent est défini comme le « mouvement horizontal orienté de l'air dans l'atmosphère terrestre ». Autrement dit, le vent est de l'air qui se déplace dans une direction déterminée. Il fait partie de notre vie quotidienne – du moins hors de nos habitations – mais nous ne nous en préoccupons guère en général.

MAIS D'OÙ VIENT-IL AU JUSTE ?

Le responsable est ici encore le soleil. Il réchauffe notre planète différemment selon la latitude. À l'équateur, où il brille à longueur d'année haut dans le ciel, le mercure descend rarement en-dessous de 30° C. Aux pôles, en revanche, le soleil ne se montre pas pendant six mois de l'année et, les six autres mois, il est si bas que les températures restent autour de 0° C. Comme la nature essaie d'équilibrer les choses, le vent véhicule l'air froid des pôles vers l'équateur, tandis que l'air chaud prend le chemin inverse.

Voilà en gros ce qu'il se passe. Dans le détail, les terres et les océans n'absorbent pas le rayonnement solaire à la même vitesse et dans les mêmes proportions. Sur les terres, l'air se réchauffe plus vite et davantage, les masses d'air se dilatent, deviennent plus légères et montent. Il en résulte

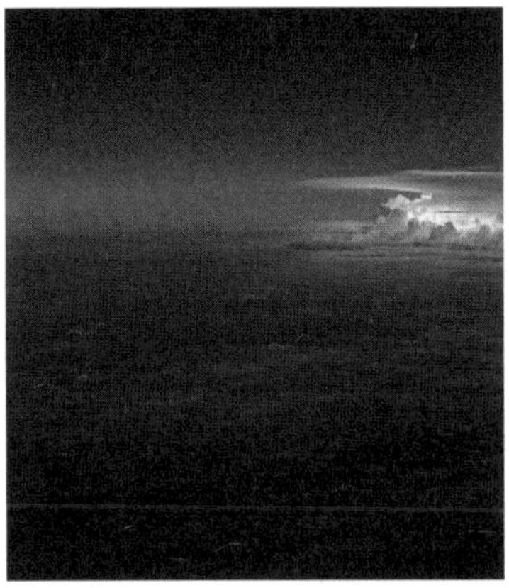 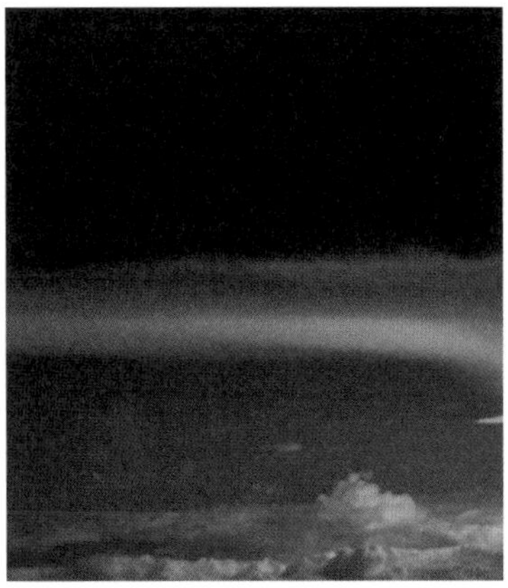 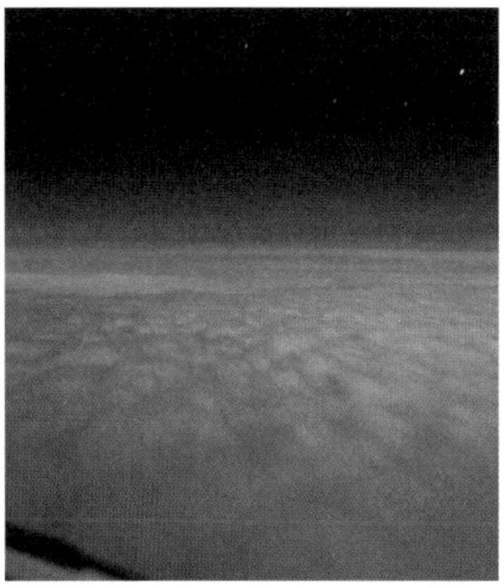

Air over land warms faster and more intensively, expanding as it does so and rising as it becomes lighter. This leads to thinner air on the ground: this is an area of low pressure. The wind counterbalances the lack of molecules by bringing in air from an area of high pressure, i.e. somewhere where too much air is competing for space. So wind always moves from areas of high pressure to areas of low pressure. The more extreme the differences between the areas, the stronger the current of air—and the greater the danger of storms.

dem Land erhitzt sich schneller und stärker, die Luftpakete dehnen sich aus, werden leichter und steigen auf. Jetzt fehlen aber am Boden an dieser Stelle Luftteilchen – es hat sich ein Tiefdruckgebiet gebildet. Der Wind kämpft nun gegen diesen Mangel an Teilchen an, indem er Luft von einem Hochdruckgebiet – wo es immer einen Luftteilchenüberschuss gibt – zum Tief transportiert. Wind weht also stets vom Hoch zum Tief. Je extremer dabei die Luftdruckgegensätze, desto stärker auch der Wind und die Gefahr von Sturmwetterlagen.

au sol une déperdition de particules d'air, une zone de basse pression. Pour compenser cette perte, le vent transporte des particules d'air d'une zone de haute pression, où elles sont toujours en excédent, vers une zone de basse pression. Le vent souffle donc toujours du trop-plein vers le déficit de particules. Si l'écart entre les pressions atmosphériques est extrême, le vent sera plus fort, et le risque de tempête accru.

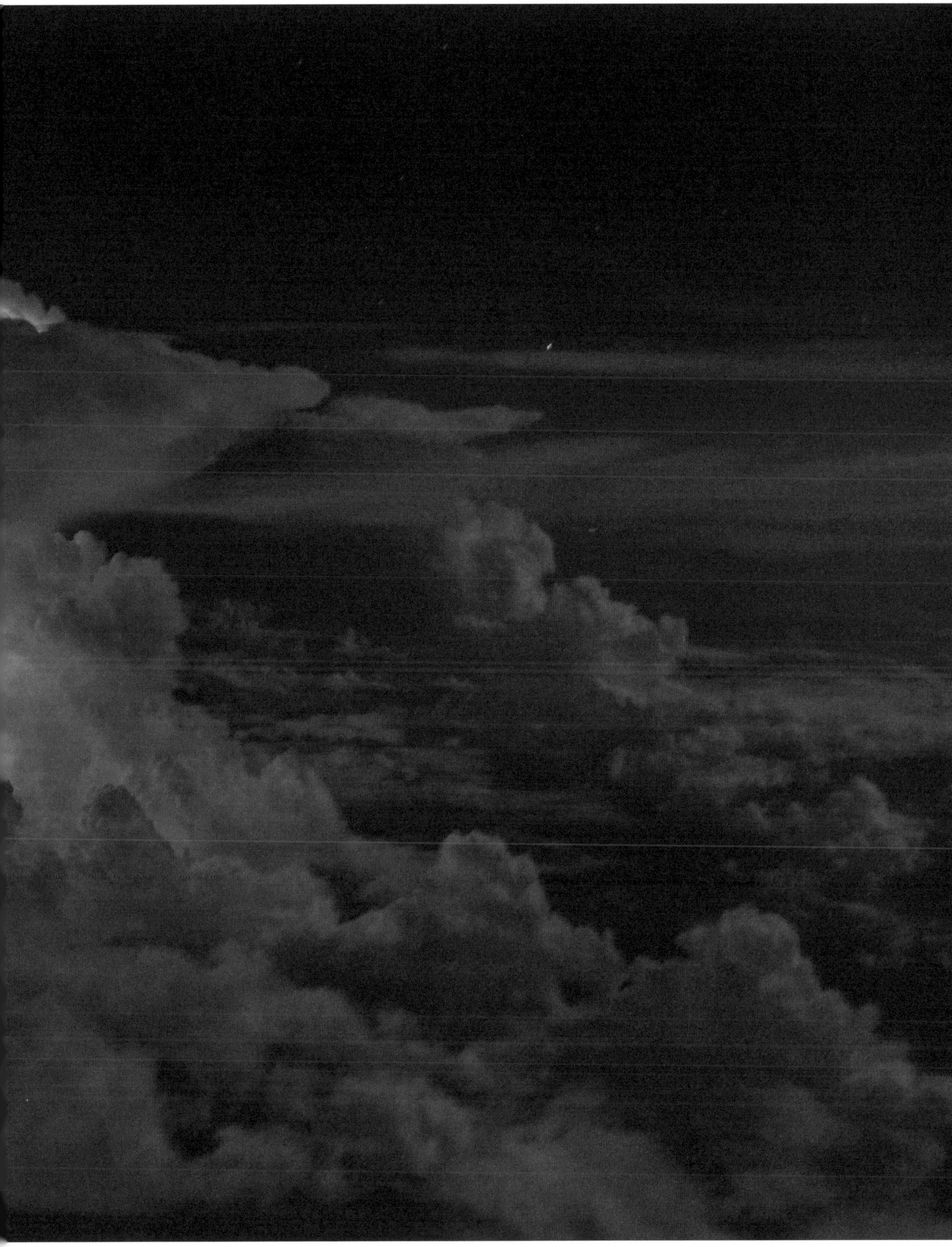

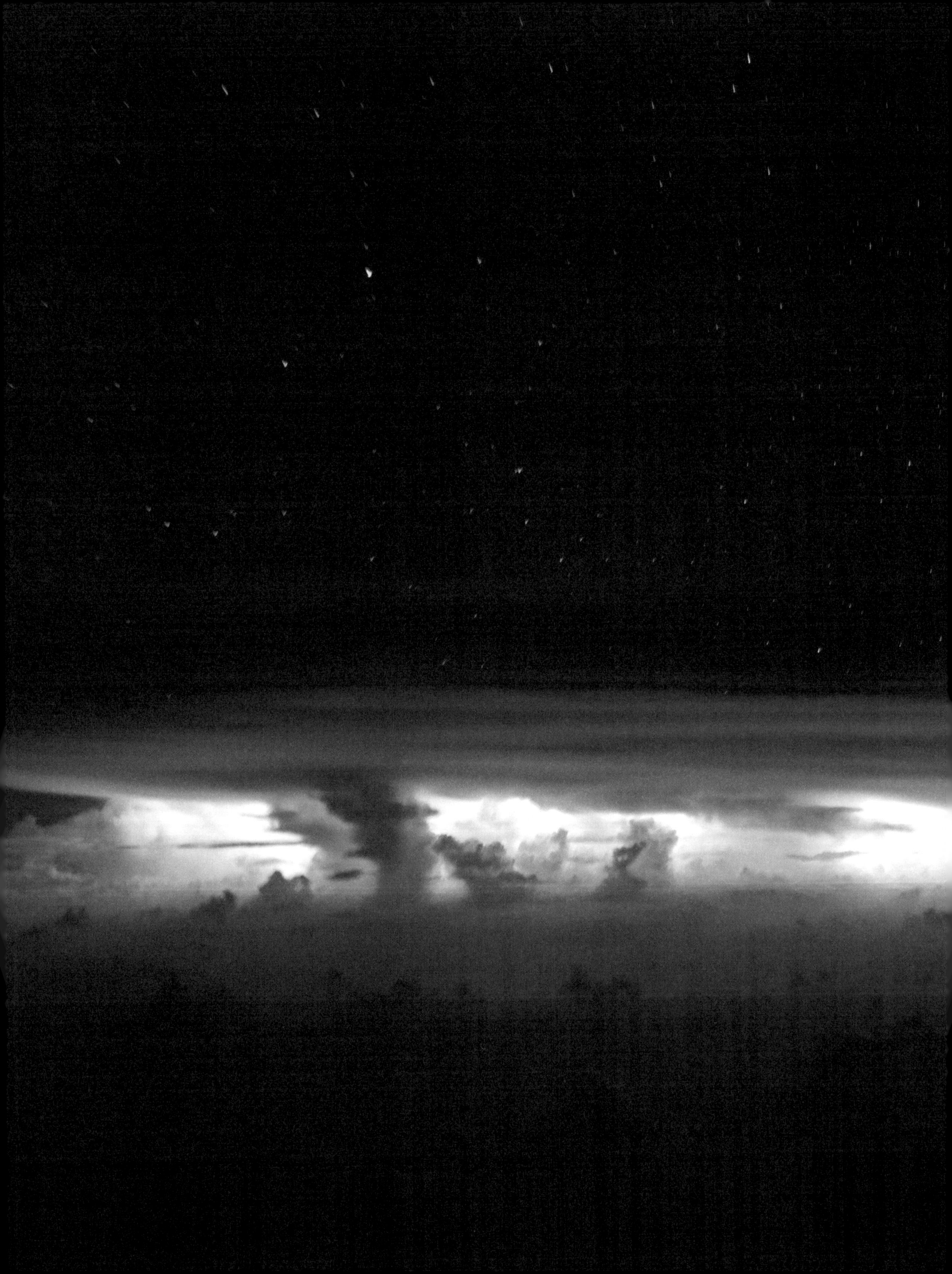

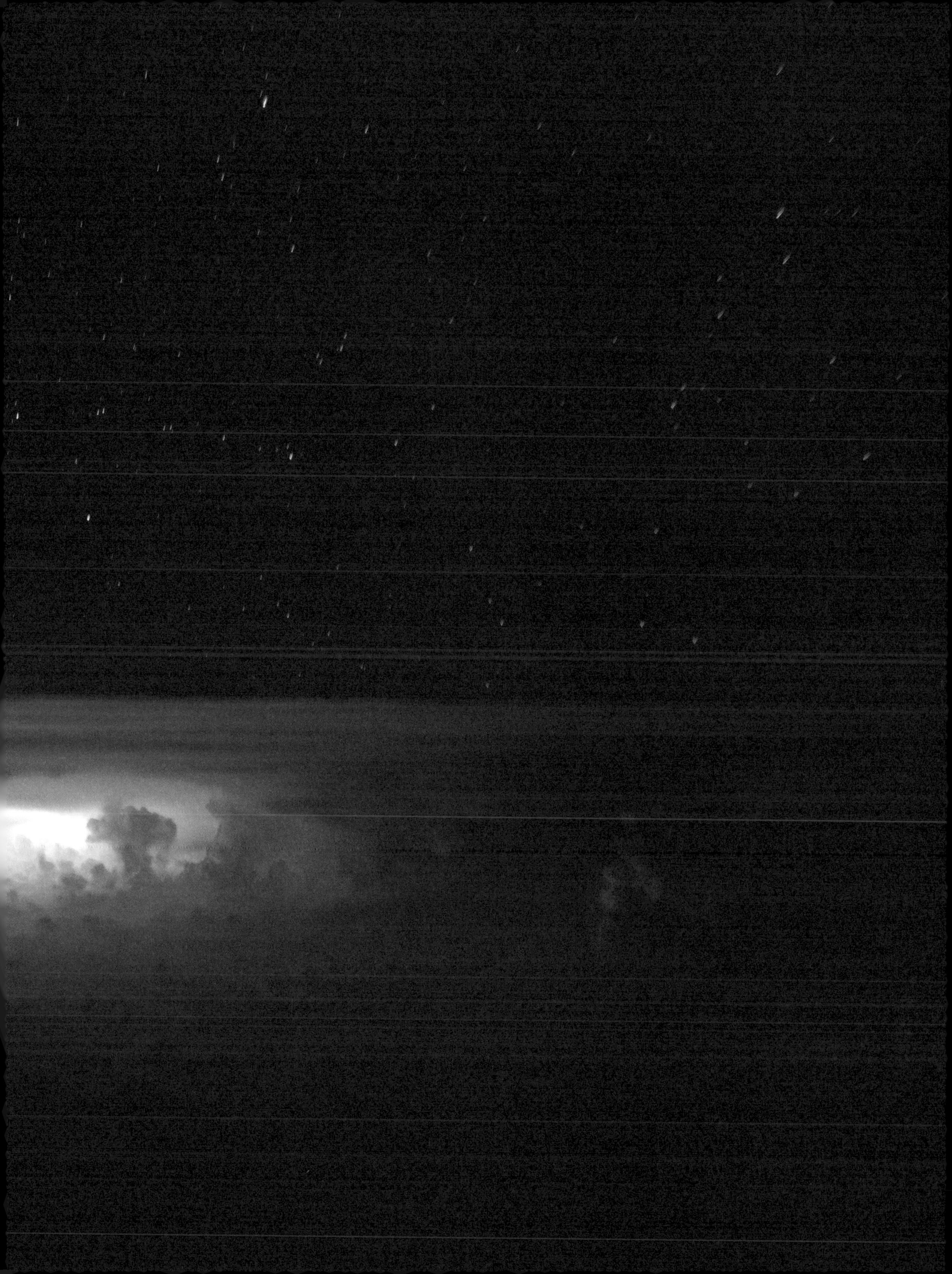

What causes thunder and lightning?

A flash, followed by a rumble and a bang, is what we see when a storm discharges electrical energy. The effects of lightning strikes at ground level can be exceptionally damaging.

BUT WHAT EXACTLY CAUSES LIGHTNING? AND WHAT HAPPENS INSIDE STORM CLOUDS TO GENERATE IT?

Simply put, the stronger the winds within a cloud, the greater the force with which droplets of water, some frozen, are thrown around; the continuous collisions between these droplets and crystals generate a huge amount of electrostatic charge, similarly to what happens between the two poles of a battery. The positively-charged ice crystals gather at the top of the cloud as the heavier, negatively-charged droplets sink to the bottom; if the tension between these two poles is high enough, this leads to a gigantic discharge of energy—and lightning happens. The bolts may travel within one cloud, between clouds, or from a cloud to the ground. Bolts of lightning are of gargantuan proportions and travel at incredible speed, covering 60,000 miles every second and heating the air around them to temperatures of up to

Wie entstehen Blitz und Donner?

Es blitzt, es donnert und kracht – ein Gewitter entlädt sich immer in Form von elektrischer Energie. Diese Energie nennen wir Blitze, und manchmal richten sie auf dem Erdboden großen Schaden an.

ABER WIE ENTSTEHT SO EIN BLITZ EIGENTLICH? UND WAS PASSIERT BEI SEINER ENTSTEHUNG IM INNEREN EINER GEWITTERWOLKE?

Je stärker die Winde in der Gewitterwolke, desto wilder werden (gefrorene) Wassertröpfchen darin herumgeschleudert. Durch das ständige Aufeinanderprallen der Teilchen und die dadurch entstehende Reibung der Tröpfchen aneinander bauen sich gewaltige elektrische Ladungen auf. Ähnlich wie bei einer Batterie entstehen entgegengesetzte Pole. Im oberen Teil der Gewitterwolke befinden sich die positiv geladenen Eiskristalle, die schweren negativ geladenen Wassertröpfchen sammeln sich unten. Ist das Spannungsfeld groß genug, gibt es einen gigantischen Kurzschluss – der Blitz ist da. Dabei können sich Blitze innerhalb einer Wolke, zwischen Wolken und zwischen Wolke und Erdboden entladen. Der Riesenfunken ist furchtbar

Comment les éclairs et le tonnerre naissent-ils ?

Éclairs, coups de tonnerre et fracas. Un orage éclate toujours en libérant de l'énergie électrique, les éclairs. Ceux-ci provoquent parfois d'importants dégâts au sol.

MAIS COMMENT UN ÉCLAIR SE FORME-T-IL AU JUSTE ? ET QUE SE PASSE-T-IL QUAND IL PREND NAISSANCE AU CŒUR D'UN NUAGE ORAGEUX ?

Plus les vents se déchaînent au sein du nuage d'orage, plus les gouttelettes d'eau (gelées) y sont brassées avec vigueur. Entrant constamment en collision, les particules d'eau se frottent les unes aux autres, engendrant des charges électriques considérables. Et le nuage devient un dipôle, à l'instar d'une pile électrique. Dans la partie supérieure du nuage d'orage se trouvent les grêlons chargés positivement, tandis que les lourdes gouttelettes d'eau chargées négativement s'accumulent dans la partie inférieure. Quand le champ magnétique est suffisamment puissant se crée un gigantesque court-circuit : l'éclair. Celui-ci peut se décharger au sein d'un nuage, entre des nuages, et entre un nuage et le sol. D'une rapidité prodigieuse (100 000 kilomètres par seconde), cette étincelle

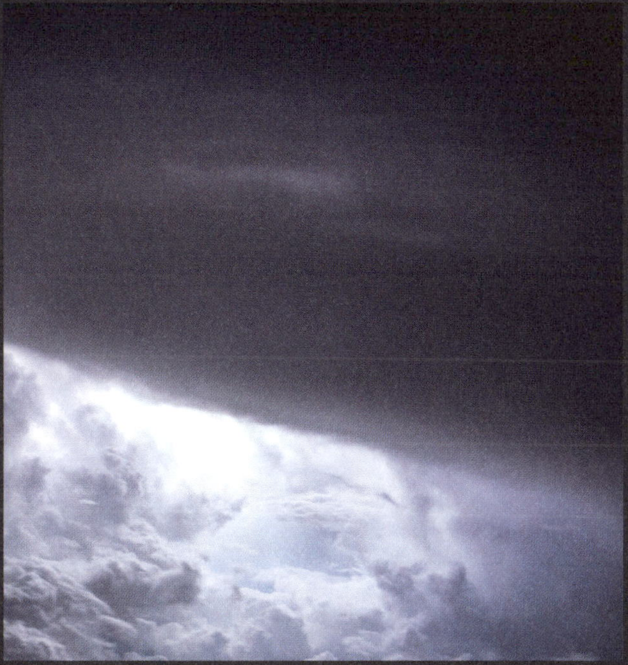

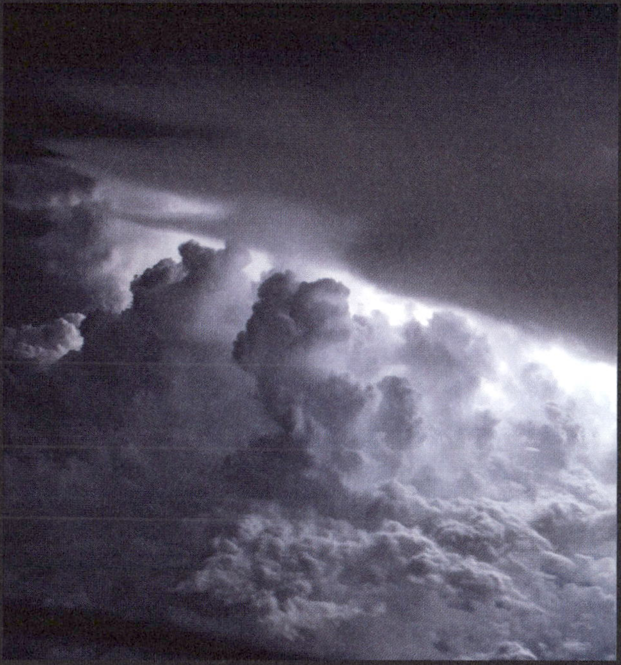

55,000°F (30,000°C). This momentary superheating causes air to expand at explosive speed—producing a shock wave, which we hear as thunder. As such, thunder and lightning occur at the same time, but because light travels at many times the speed of sound, we usually see lightning before we hear thunder.

The discharge channel of lightning is only a few centimetres in diameter, but within it, several million volts of electrical tension can occur, generating more than 100,000 amperes of electrical current. As air is a weak conductor, however, and because lighting seeks the path of least electrical resistance, it forks repeatedly or travels in characteristically unpredictable zig-zags.

schnell: In einer Sekunde legt er 100.000 Kilometer zurück und er erhitzt die Luft bis auf 30.000 Grad Celsius. Durch die plötzliche Hitze dehnt sich die Luft explosionsartig aus. Bei diesem Prozess entsteht ein lauter Knall – der Donner. Blitz und Donner entstehen zwar immer gleichzeitig, doch ist der Blitz oft früher zu sehen, da sich Licht schneller ausbreitet als der Schall.

Die Blitzbahn hat einen Durchmesser von wenigen Zentimetern. Es wird darin aber extrem viel Spannung, nämlich mehrere Millionen Volt, erzeugt. Die Stromstärke kann mehr als 100.000 Ampere betragen. Da Luft Strom schlecht leitet und der Blitz Gebiete mit großem elektrischem Widerstand umgeht, verästeln sich Blitze immer wieder, oder bilden seltsame Zackenlinien.

géante élève la température de l'air à jusqu'à 30 000 degrés Celsius. Sous l'effet de cette fournaise subite, l'air se dilate en produisant un bruit d'explosion, le tonnerre. L'éclair et le tonnerre sont toujours synchrones mais, la lumière se propageant plus vite que le son, on perçoit souvent l'éclair en premier.

La trajectoire de l'éclair est large de quelques centimètres seulement, mais la tension qui s'y développe atteint des valeurs faramineuses de plusieurs millions de volts. L'intensité du courant peut dépasser les 100 000 ampères. Comme l'air conduit mal l'électricité et que l'éclair contourne les zones présentant une résistance électrique élevée, les éclairs se ramifient ou dessinent de curieux zigzags.

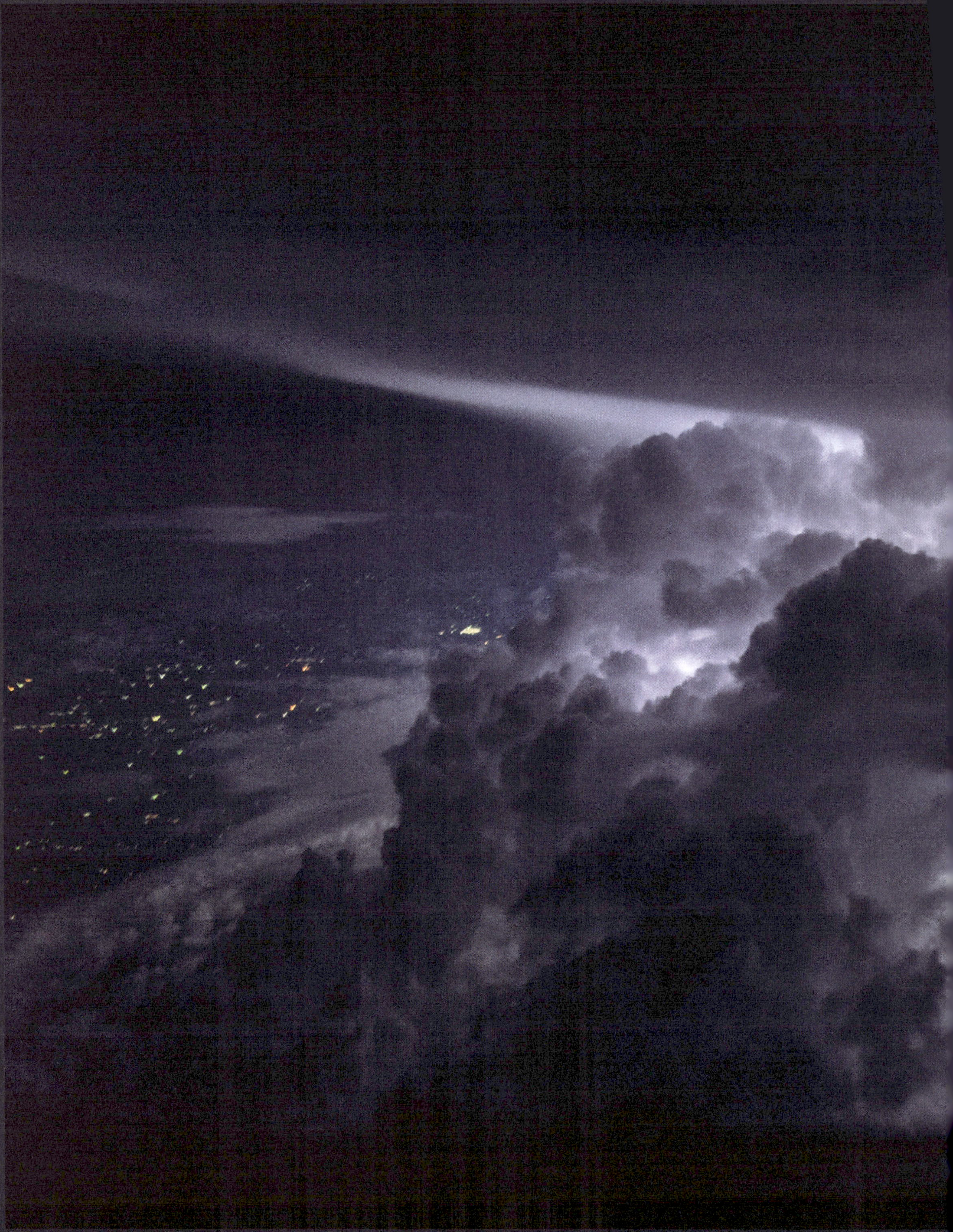

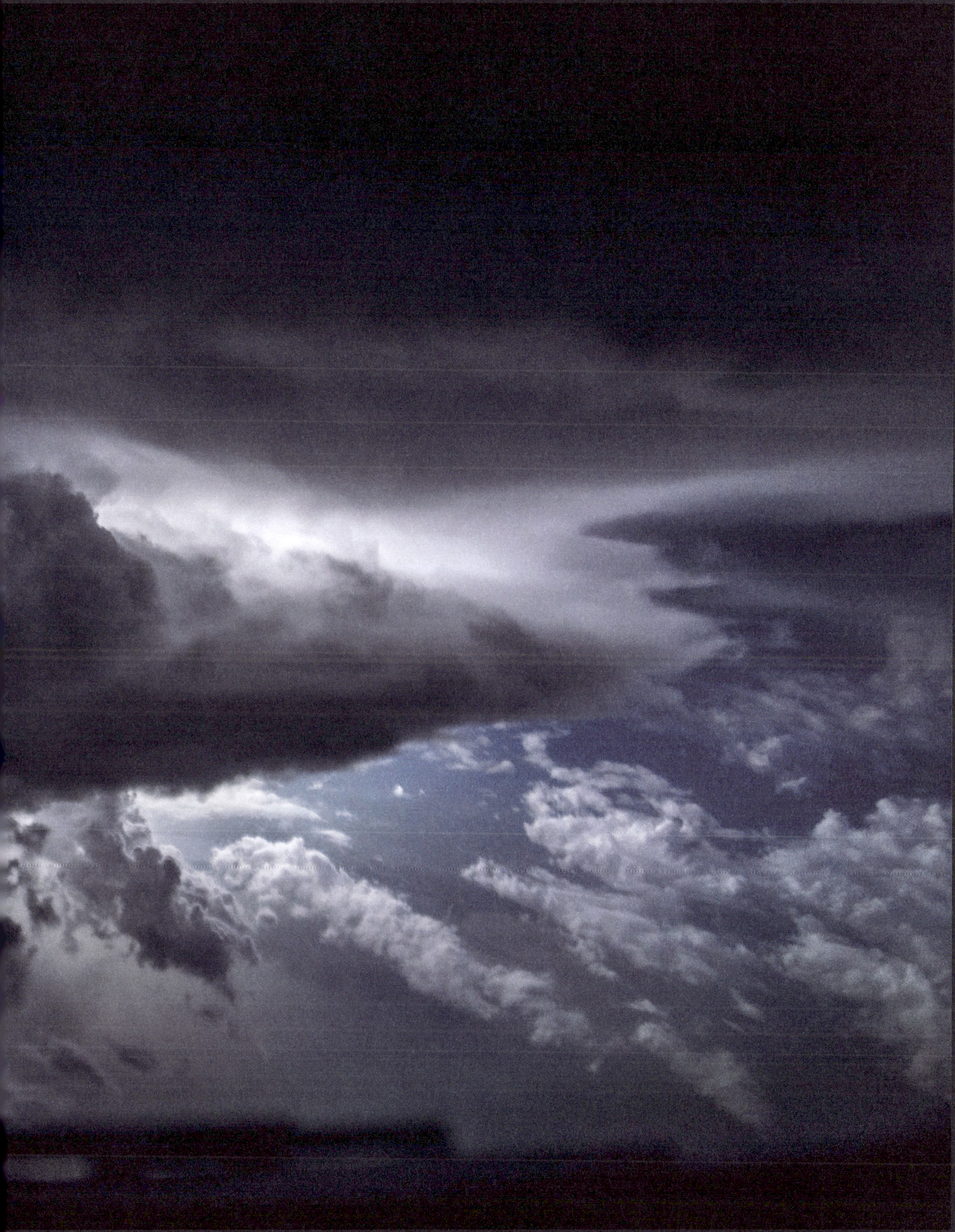

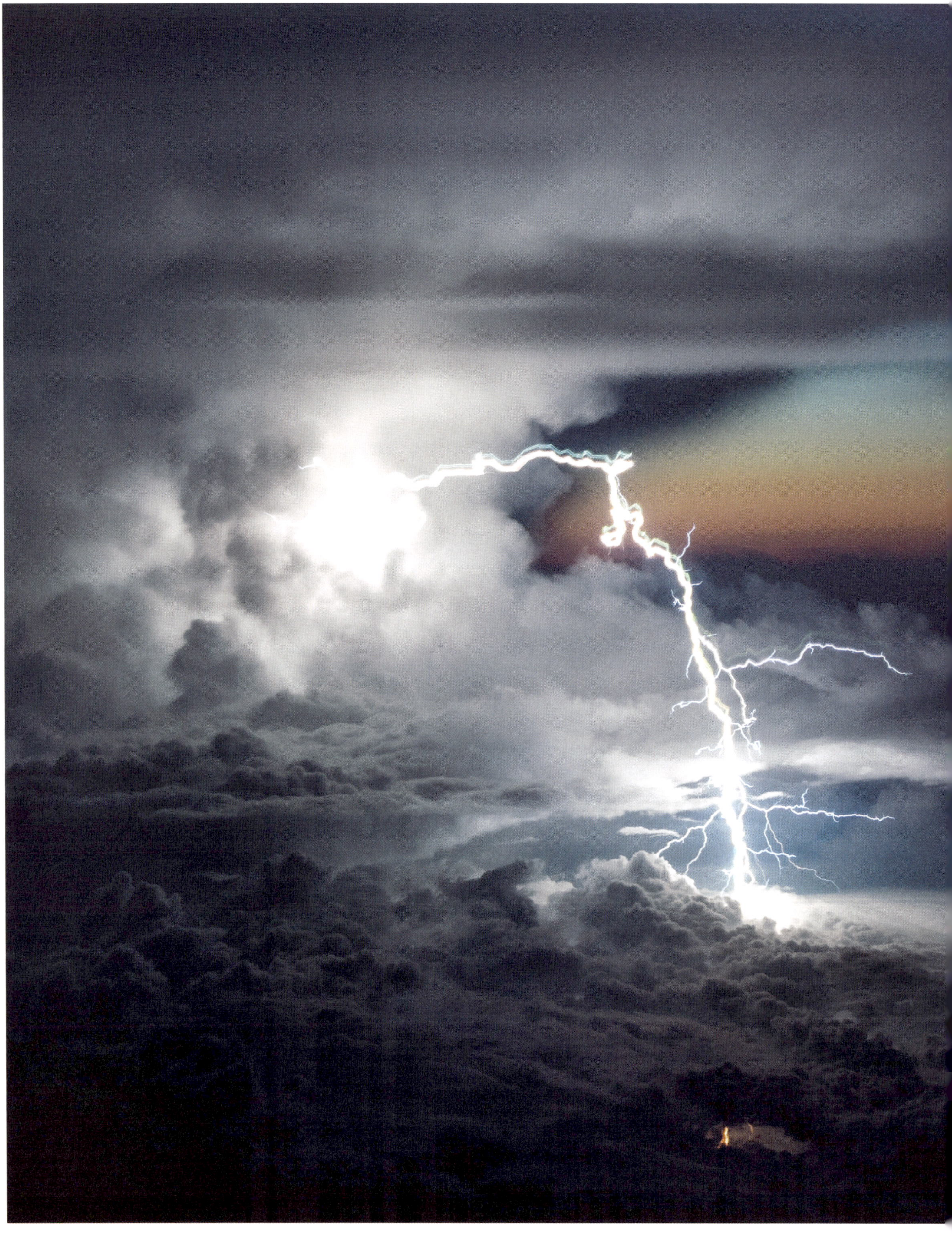

How often does lightning happen?

During a typical summer storm, thunder and lightning generally last for around half an hour. If the storm is not directly above the person watching it and in the absence of strong winds or heavy downpours, the phenomenon is somewhat unspectacular. Nature, however, has a few tricks up its sleeve…

In temperate latitudes, for instance, cold fronts in summer bring areas of air with vastly different temperatures into contact, creating storms, which are both extremely dangerous and exceptionally lightning-intensive, with up to 10,000 bolts in 30 minutes occurring regularly in Europe. On a global level, there are around 1,000 storm events daily which produce between four and ten million bolts of lightning. That means that, on average, there are around 50 lightning strikes per second. Some of these hit trees or buildings, causing considerable damage; all of them produce a big bang.

Thunderstorms are most frequent in the tropics, where the atmosphere has constant levels of the humidity and warmth necessary for serious storm conditions. Lake Maracaibo in Venezuela is the lightning capital of the world, with around 232 strikes per year, per square kilometre; with 0.6 to 5 strikes annually, the average square kilometre of European landmass is comparatively quiet, but of course the lightning season only lasts from May to September, when the weather is warm enough. In equatorial regions, the consistently high temperatures mean that heavy thunderstorms can occur at any time of year.

Wie oft blitzt es?

Bei einem Sommergewitter blitzt und donnert es meist für nur relativ kurze Zeit. Wenn das Gewitter sich nicht direkt über dem Betrachter entlädt, kein heftiger Wind aufbraust und es nicht regnet wie aus Eimern, erscheint es allerdings eher unspektakulär. Die Natur kann aber auch ganz anders …

In den mittleren Breiten sind es vor allem Gewitter an Kaltfronten im Sommer, dort, wo große Temperaturunterschiede aufeinandertreffen, die kreuzgefährlich und sehr blitzintensiv werden können – 10.000 Blitze in einer halben Stunde über Deutschland sind da keine Seltenheit. Weltweit entstehen täglich etwa 1000 Gewitter mit etwa vier bis zehn Millionen Blitzen. Im Mittel sind das pro Sekunde etwa 50 Blitzeinschläge. Blitze also, die den Boden, Bäume oder Gebäude berühren, und dabei erheblichen Schaden anrichten können – und in jedem Fall einen ordentlichen Knall produzieren.

Die meisten Gewitter gibt es in den Tropen. Hier sind genug Feuchtigkeit und Wärme in der Atmosphäre – das Futter für jedes gefährliche Unwetter. Dabei ist der Maracaibo-See in Venezuela der blitzreichste Ort der Erde – mit etwa 232 Blitzen pro Jahr und Quadratkilometer. Da geht es in Mitteleuropa mit 0,6 bis 5 Blitzen eher ruhig zu. Allerdings ist in Europa auch nur in den wärmeren Monaten zwischen Mai und September Blitzsaison, während die Tropen – bedingt durch die durchgehend warmen Temperaturen – das ganze Jahr über mit heftigen Gewittern rechnen müssen.

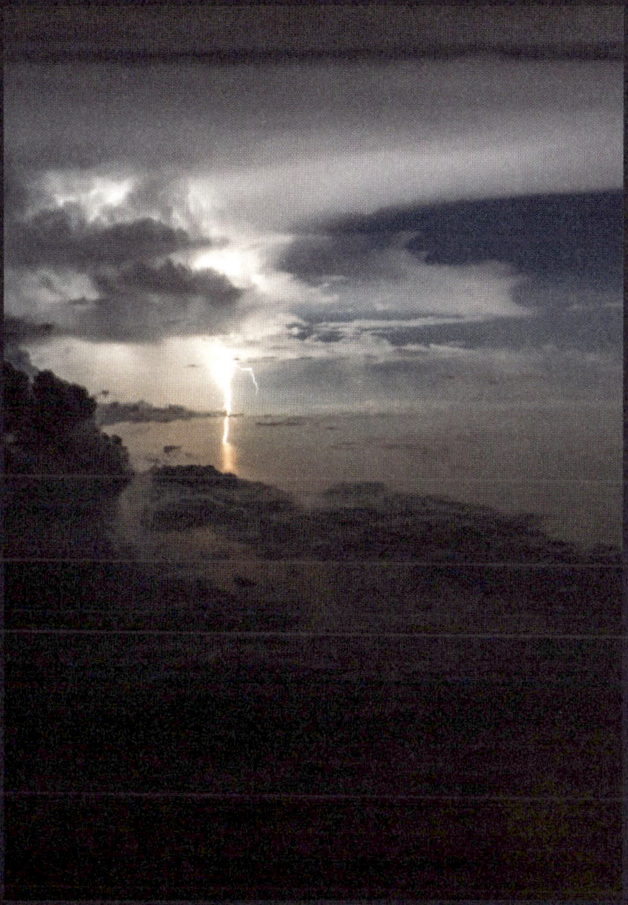

À quelle fréquence les éclairs se produisent-ils ?

Pendant un orage d'été, il y a en général des éclairs et du tonnerre pendant environ une demi-heure. Quand l'orage n'éclate pas juste au-dessus de l'observateur, qu'aucun vent violent ne souffle et qu'il ne pleut pas à verses, cela n'a rien de spectaculaire. Mais la Nature réserve des surprises...

Sous les latitudes moyennes, les orages surviennent avant tout en été sur les fronts froids, où les écarts de température sont importants. Ils peuvent être mortellement dangereux et s'accompagner d'un nombre incroyable d'éclairs : il n'est pas rare que 10 000 éclairs zèbrent le ciel allemand en l'espace d'une demi-heure. À l'échelle de la planète, on enregistre chaque jour environ 1 000 orages, qui s'accompagnent de 4 à 10 000 millions d'éclairs. Cela représente en moyenne par seconde environ 50 impacts de foudre, c'est-à-dire éclairs qui, dans un fracas effroyable, touchent le sol, des arbres ou des édifices, et peuvent ce faisant causer des dégâts considérables.

La plupart des orages éclatent dans la zone intertropicale. L'humidité et la chaleur ambiantes, qui alimentent ces perturbations dangereuses, y sont propices. Le lac Maracaibo, au Venezuela, est le site le plus foudroyé au monde, avec environ 232 impacts de foudre par an au kilomètre carré. L'Europe centrale, où cette valeur est comprise entre 0,6 et 5 impacts, est une région plutôt calme. Précisons que les éclairs ne frappent l'Europe que pendant les mois chauds, de mai à septembre, tandis que sous les tropiques, où il fait chaud en permanence, des orages violents sévissent à longueur d'année.

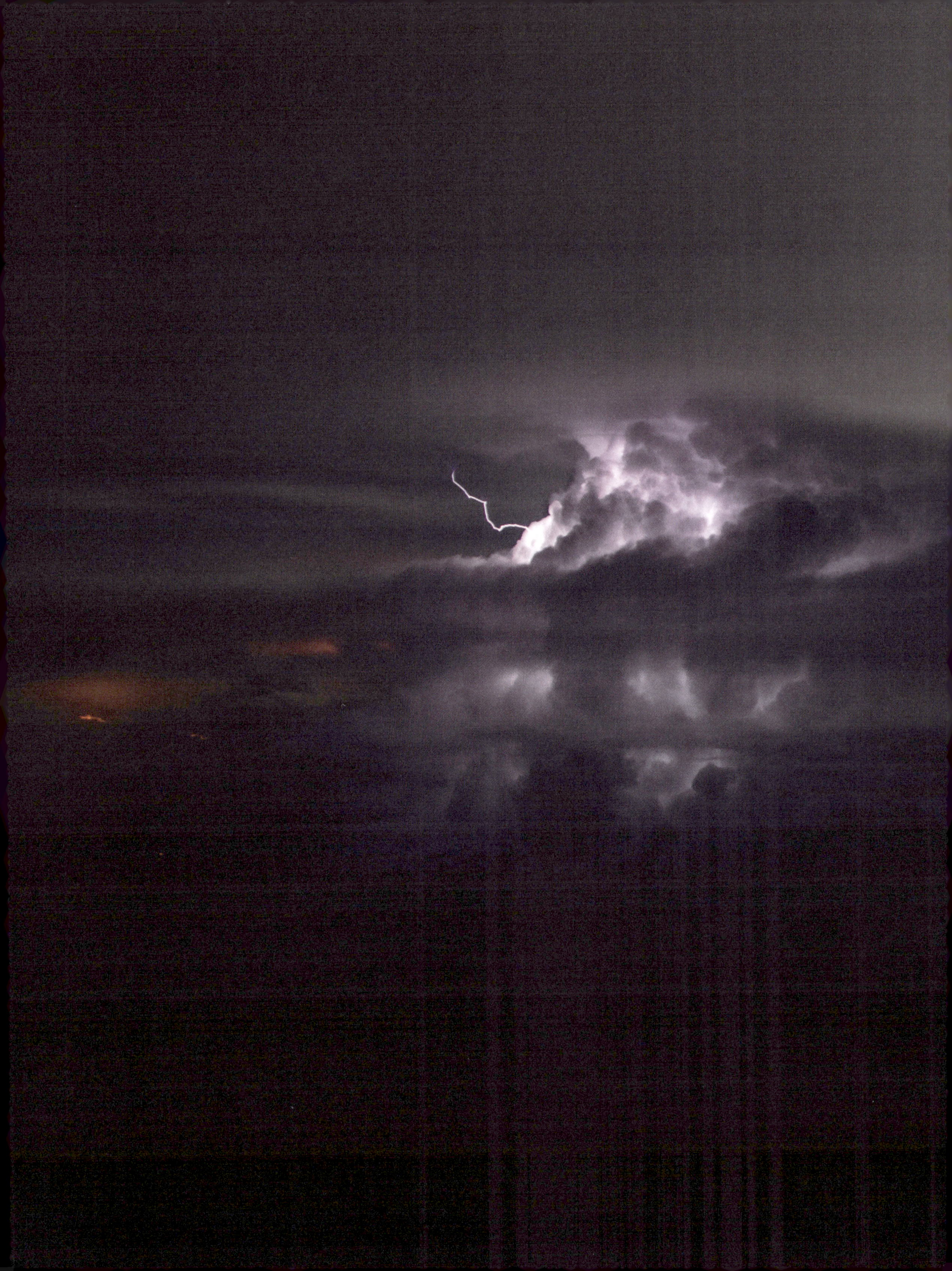

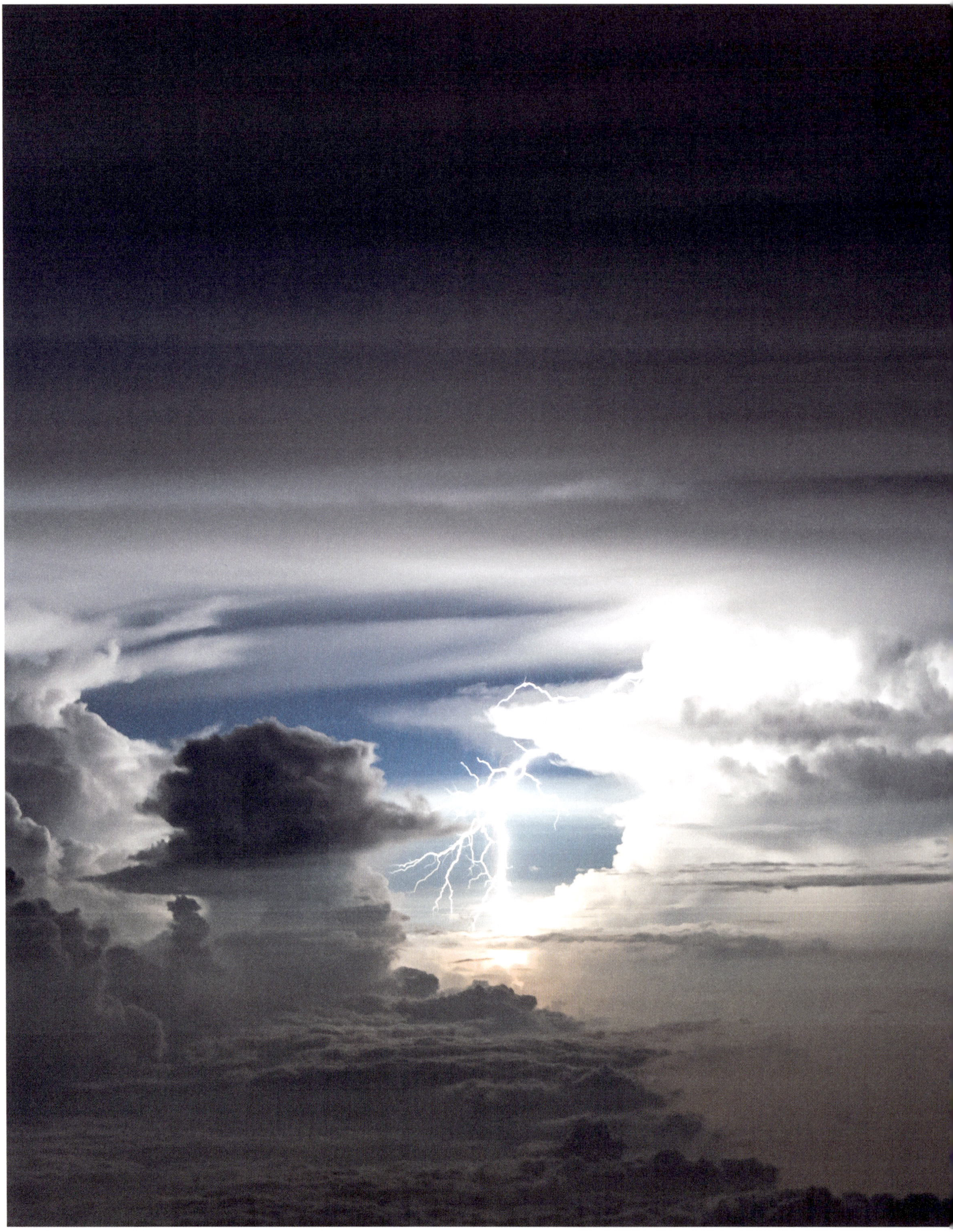

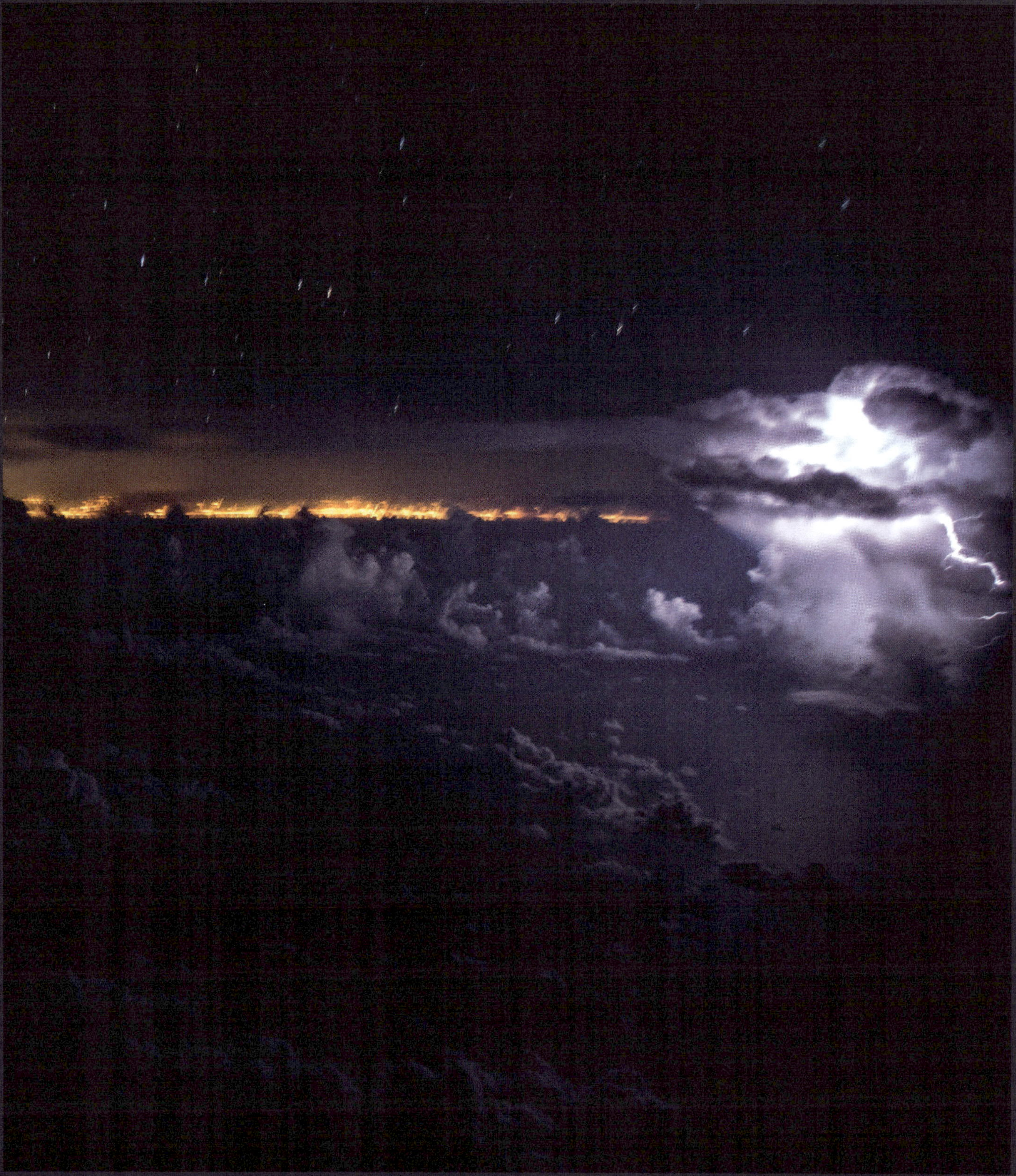

A #lightning is so powerful it lights up dozens of kilometers around it at once

How long are bolts of lightning?

When lightning shoots from a cloud in the night sky down to the ground, forking out and momentarily bathing the heavens in a glaring flash, the visual effect is impressive. In terms of length, these cloud-to-ground bolts vary, from 3,000 to 7,000 feet in length in temperate latitudes through to anything up to 10,000 feet in the tropics, where humidity is higher. Bolts of lightning, which jump from cloud to cloud, are usually longer, spanning between 16,000 to 23,000 feet. The longest bolts are created when the discharge occurs in the upper part of a storm cloud and goes to ground, snaking all the way down through the cloud and reaching lengths of up to 33,000 feet. Nevertheless, every now and then, nature likes to make a mockery of standards and averages: the longest ever bolt of lightning sighted occurred over Oklahoma in 2007, jumping 200 miles across the heavens—or, in European dimensions, from Brussels to Paris.

Welche Länge haben Blitze?

Es sieht schon beeindruckend aus, wenn ein Blitz nachts von einer Wolke Richtung Erdboden schießt, sich dabei verästelt, und der ganze Himmel in grellem Licht zuckt und aufleuchtet. Solche Erdblitze können unterschiedlich lang sein, in den mittleren Breiten meist ein bis zwei Kilometer, in den Tropen durch die höhere Luftfeuchtigkeit schon mal bis zu drei Kilometer. Wolkenblitze, Blitze also, die sich zwischen den Wolken entladen, sind meist fünf bis sieben Kilometer lang; Blitze, die im oberen Teil einer Gewitterwolke entstehen und sich – durch das Wolkenungetüm hindurch – bis zum Erdboden erstrecken, können eine Länge von bis zu zehn Kilometern erreichen. Manchmal allerdings setzt sich die Natur hinweg über jedes Maß und jeden Richtwert: Der längste bisher vermessene Blitz wurde 2007 über Oklahoma mit einer horizontalen Länge von 321 Kilometern gesichtet. Das ist die unglaubliche Entfernung von Brüssel bis Paris.

Combien les éclairs mesurent-ils ?

Il est vraiment très impressionnant de voir de nuit un traceur surgi d'un nuage se ramifier en direction du sol tandis que le ciel entier tressaille et s'illumine d'une lumière vive. Ces éclairs qui touchent le sol ont une longueur variant de 1 à 2 kilomètres sous les latitudes moyennes, et de 2 à 3 kilomètres sous les tropiques, où l'humidité atmosphérique est plus élevée. Les éclairs qui se déchargent entre des nuages mesurent en général de 5 à 7 kilomètres de long ; les éclairs qui prennent naissance dans la partie supérieure d'un nuage orageux et qui se frayent un passage à travers les masses nuageuses jusqu'au sol peuvent atteindre une longueur de 10 kilomètres. Mais parfois, la Nature fait fi de toute mesure et valeur indicative : l'éclair le plus long enregistré à ce jour a été observé dans le ciel de l'Oklahoma en 2007. Il atteignait à l'horizontale la longueur incroyable de 321 kilomètres, soit la distance séparant Paris de Bruxelles.

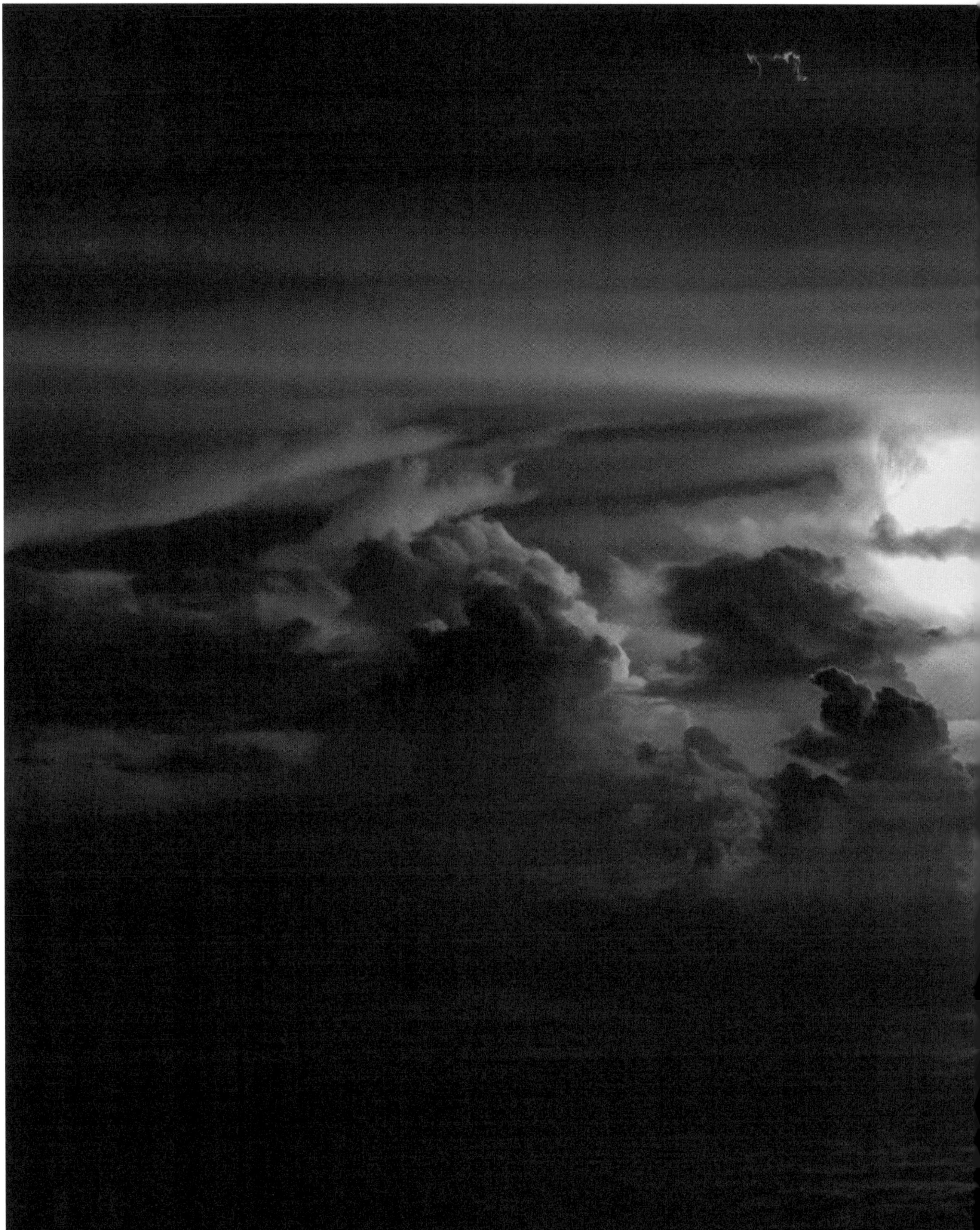

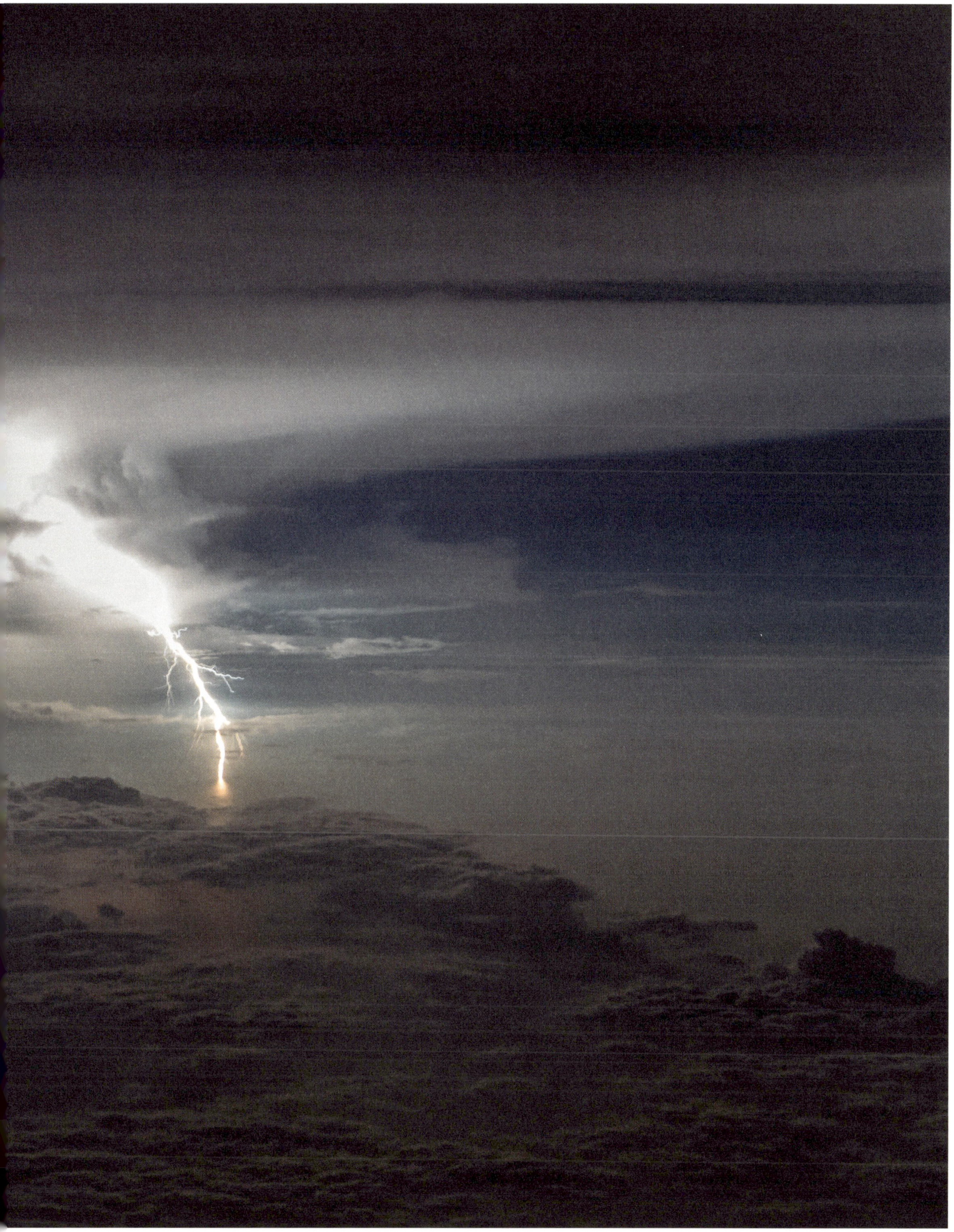

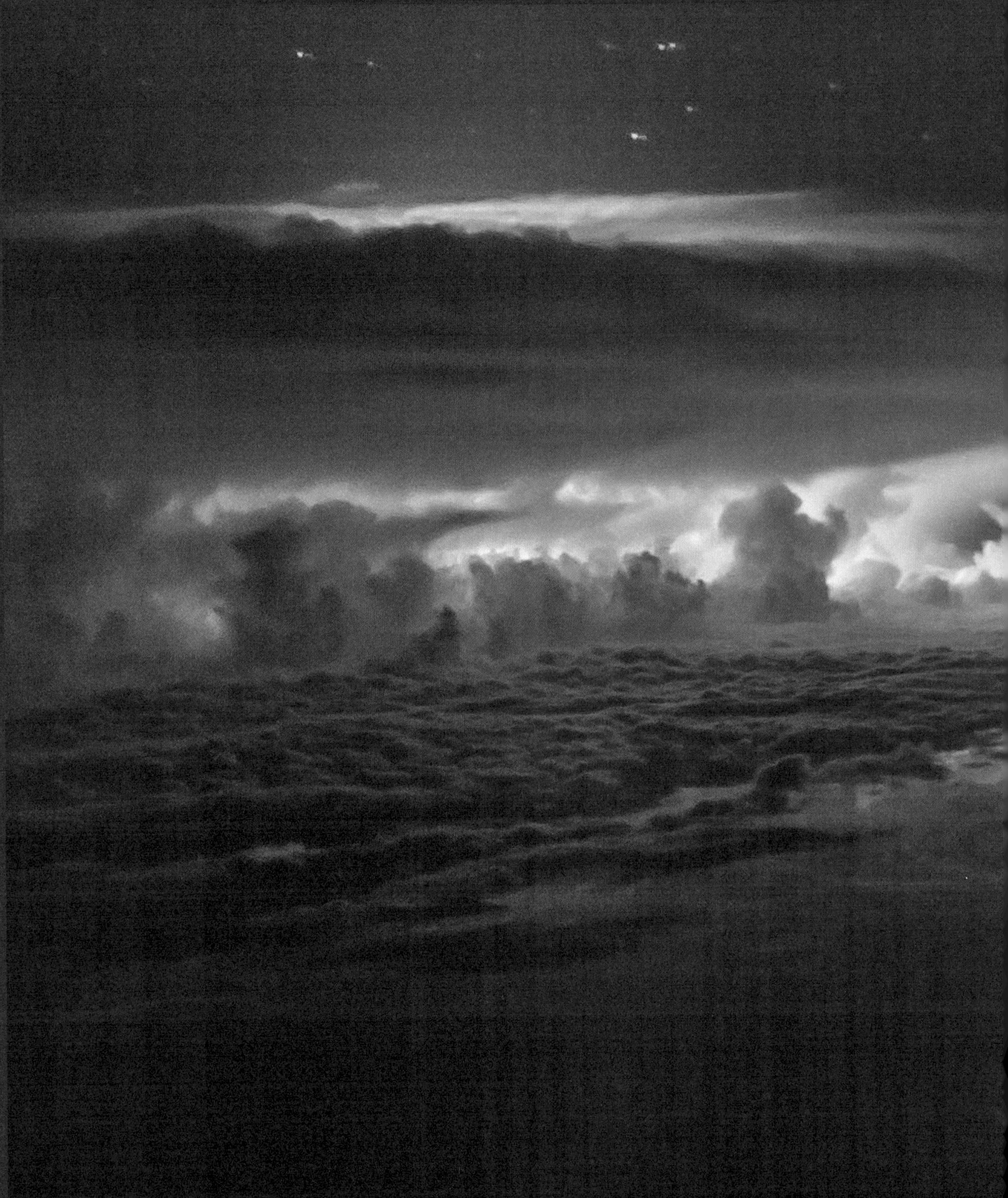

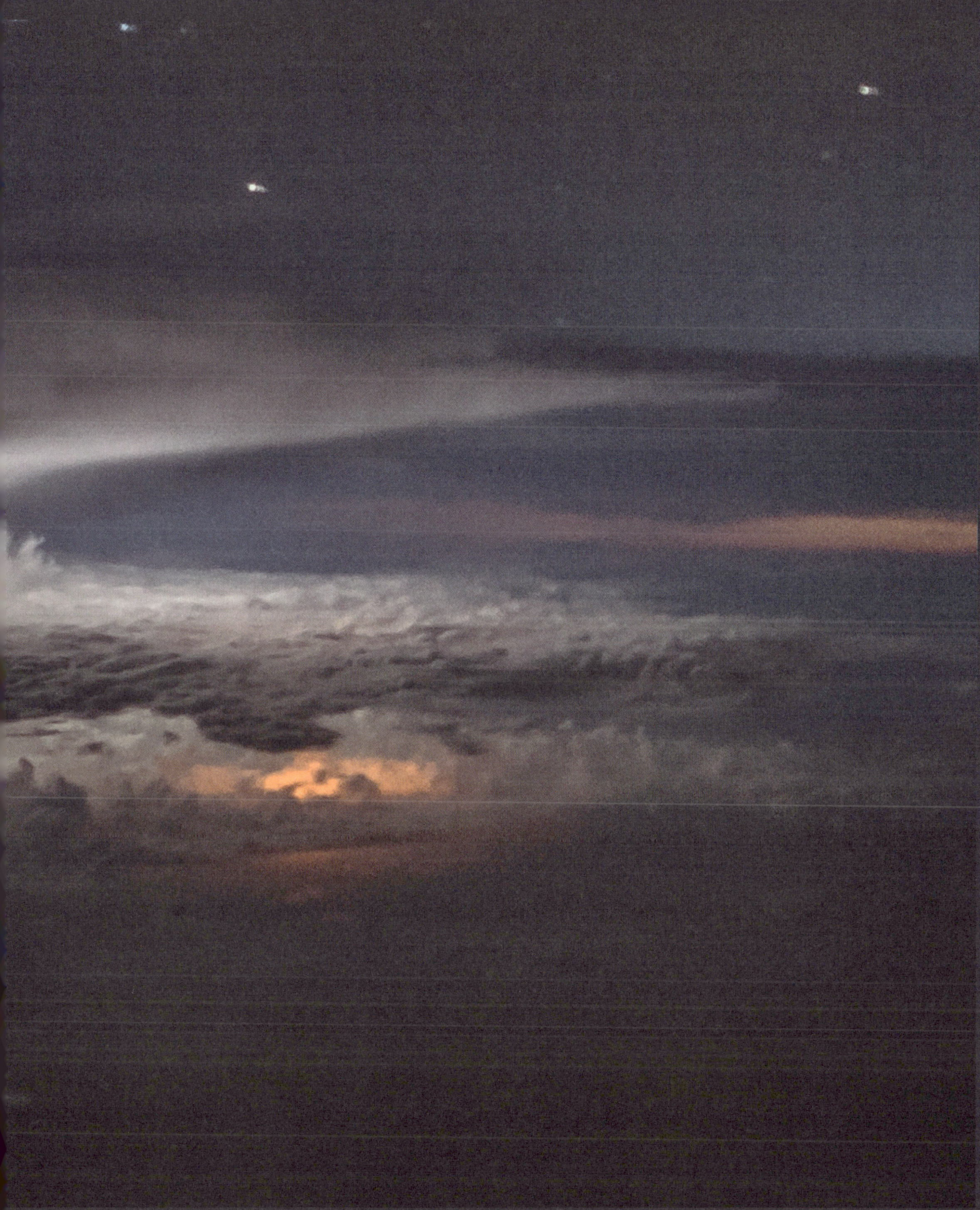

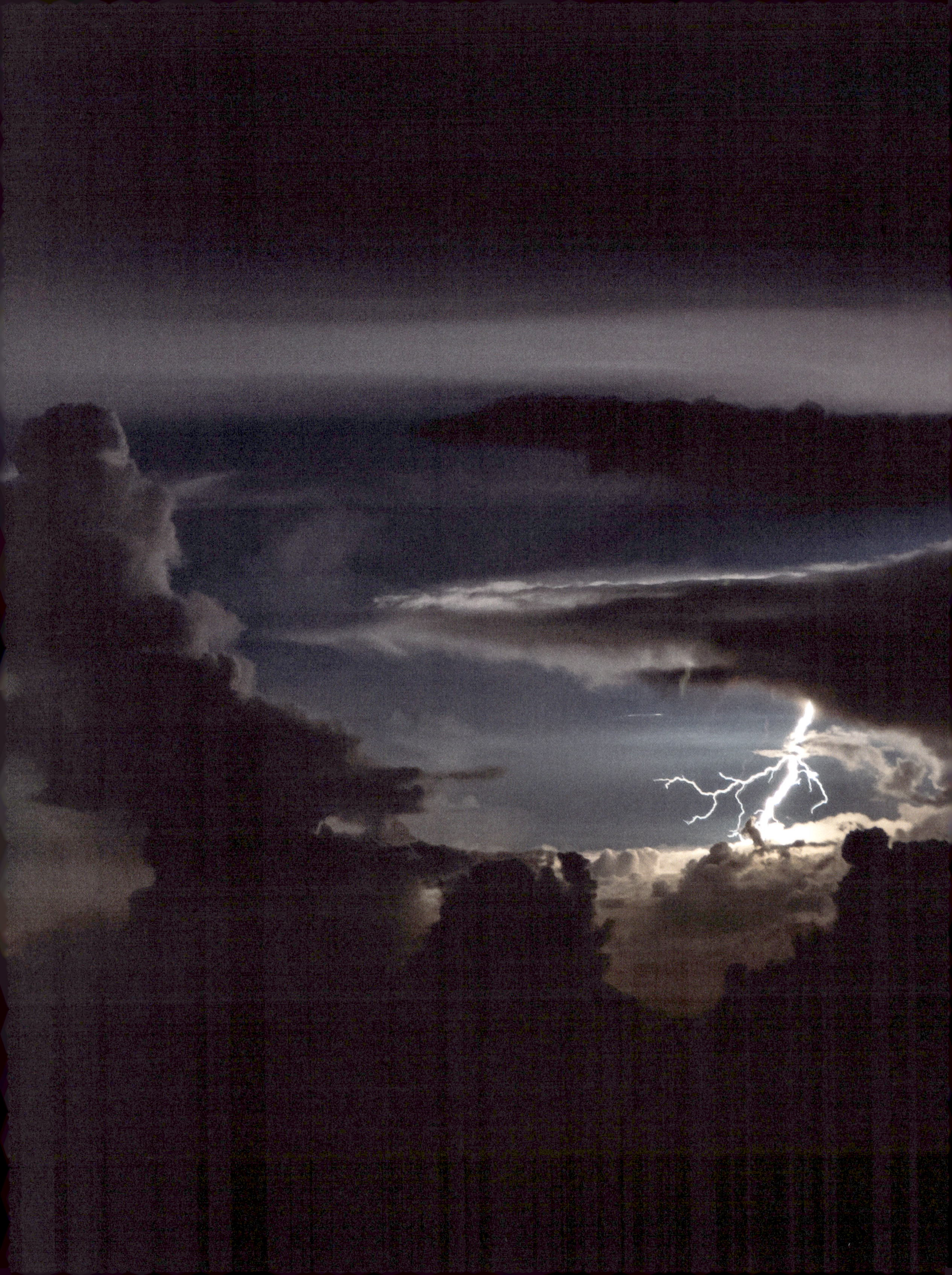

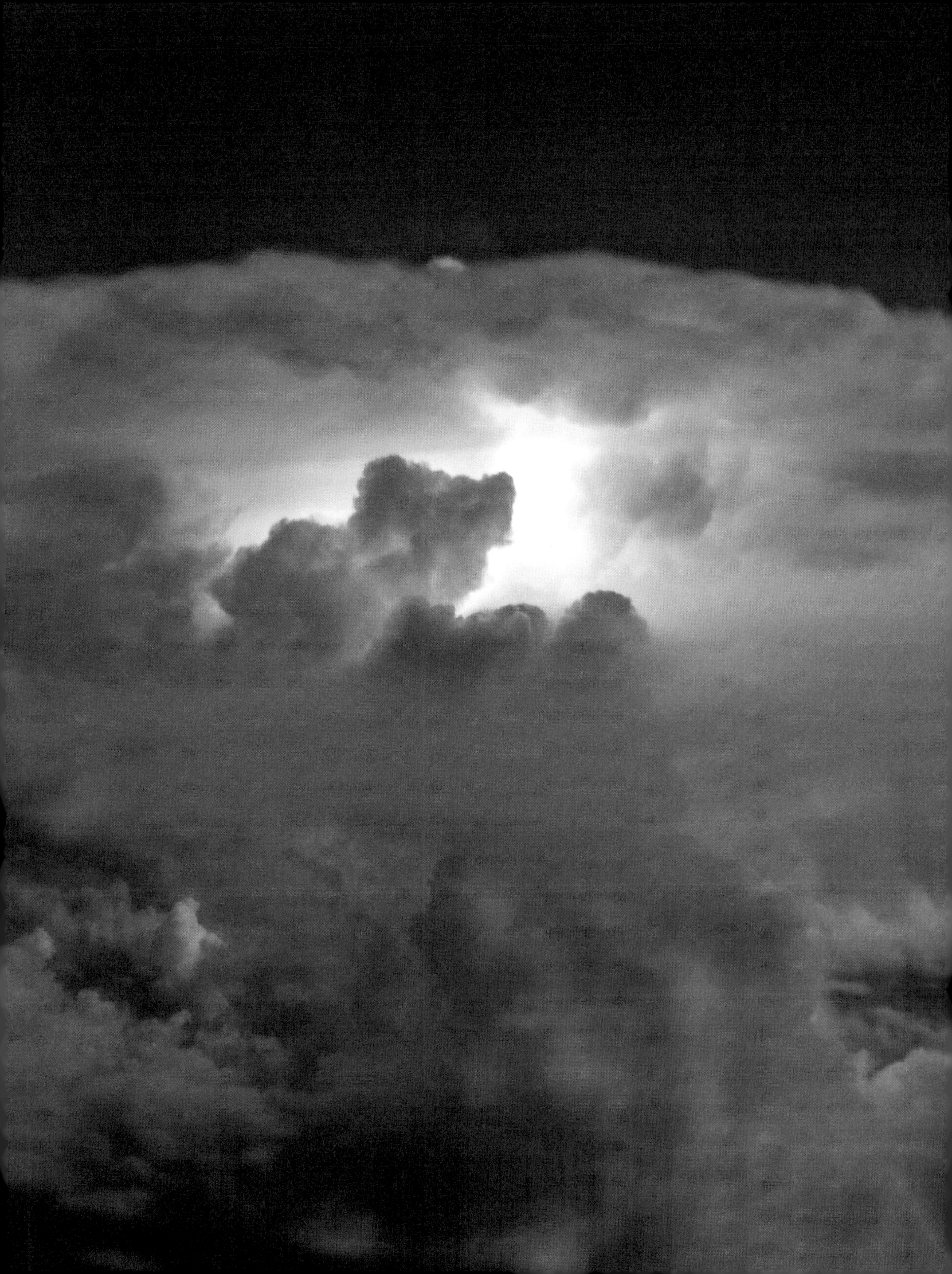

#WhyIFly

#We all learn about #lift and #drag,
but deep inside we #pilots know that it's
all #magic. And once up there, everything
else is moved and done by more of
this #magic

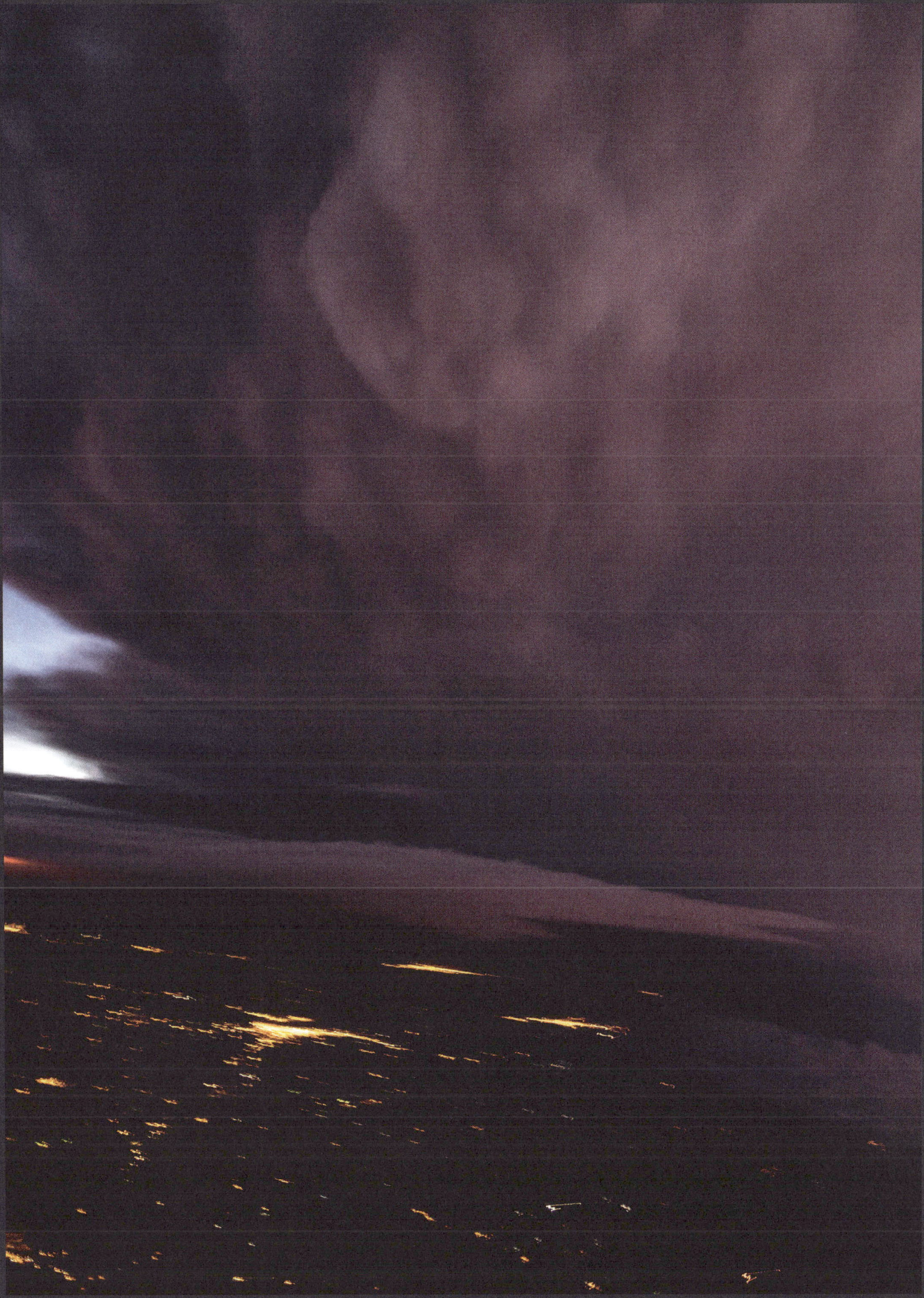

What happens when lightning strikes?

A storm is drawing in and suddenly, a bolt of lightning strikes a birch; a man stood near the tree collapses. When the ambulance crew arrives, he is beyond resuscitation—despite the fact that he shows no outward signs of injury. Lightning can be deeply sinister.

Everyone knows that lightning is dangerous, but few know just how much power a bolt of lightning generates. Even those who survive being struck by lightning generally suffer life-changing injuries—and with 20,000 amps of electric current in play, it's hardly surprising that they do.

Lightning strikes can cause serious damage to buildings too, even if they don't take a direct hit; electrical equipment is susceptible to damage if lightning strikes within a radius of one mile. If a bolt of lightning strikes a house directly, the electricity tends to travel through water pipes and electrical wiring down to the ground. As it does, the immense amount of energy released and high temperature of the bolt can lead to fires, usually at roof level; all electrical systems are generally destroyed. Despite this, people are best protected from the effects of lightning inside buildings.

Was passiert beim Blitzeinschlag?

Ein Gewitter zieht heran, und schon schlägt ein Blitz in eine Birke ein. Ein Mann, der ganz in der Nähe steht, bricht zusammen, der Notarzt kann nur noch seinen Tod feststellen. Dabei sind beim Betroffenen keinerlei äußere Verletzungen festzustellen – gruselig.

Dass Blitze gefährlich sein können, weiß jedes Kind. Aber was für Kräfte sie wirklich freisetzen, ist den Wenigsten klar. Selbst wer einen Blitzschlag in den eigenen Körper überlebt, hat meist bleibende Schäden ein Leben lang. Bei Stromstärken von über 20.000 Ampere ist das auch kein Wunder.

Auch an Häusern kann schwere Zerstörung auftreten, selbst wenn das Haus nicht direkt vom Blitz getroffen wurde. Ein Einschlag im Umkreis von zwei Kilometern reicht, um Schäden an Elektrogeräten zu verursachen. Trifft der Blitz ein Haus direkt, fließt der Strom hauptsächlich durch die Wasser- und Stromleitungen zur Erde ab. Durch die große Energie und die hohen Temperaturen im Blitz kann es zu Hausbränden kommen, meist gerät der Dachstuhl in Brand. Außerdem ist häufig die komplette Stromanlage des Hauses zerstört. Als Mensch ist

Que se passe-t-il lors d'un impact de foudre ?

Alors qu'un orage approche, un éclair foudroie un bouleau, et un homme qui se trouve dans les parages s'effondre. Les secours ne peuvent que constater le décès. Chose effroyable, la victime ne présente aucune blessure apparente.

La foudre est dangereuse, comme chacun sait. Mais peu de personnes savent combien sa décharge électrique est violente. Si l'on peut parfois survivre au foudroiement, on en garde en général des séquelles à vie. Avec des courants de plus de 20 000 ampères cela n'a rien d'étonnant.

Un bâtiment peut lui aussi subir des dégâts considérables, même sans avoir été frappé directement par la foudre. Un impact de foudre dans un rayon de 2 kilomètres suffit pour endommager des appareils électriques. Si la foudre tombe directement sur une maison, l'électricité chemine à travers les conduites d'eau et d'électricité avant de s'enfoncer dans le sol. En raison de sa charge électrique formidable et de ses températures élevées, la foudre peut provoquer un incendie dans la maison, notamment de la charpente, ou encore en détruire toute l'installation électrique. C'est cependant à l'intérieur d'un bâtiment

Lightning very rarely strikes planes, whose pilots use radar to locate and avoid storms by a generous margin. In the rare cases in which it does happen, the metal exterior of an aircraft functions as a lightning rod and given that an airborne plane is not earthed, the lightning bolt leaves as quickly as it hits. As a general rule however, several other bolts follow, and most planes are hit three to five times—some have recorded up to 25 strikes. Inside the aircraft, a low-pitched bang is often heard. Since aircrafts are essentially Faraday cages, there is no danger to the people inside: the bolt of lightning is conducted along the fuselage and continues on its way. Nevertheless, aircrew document lightning strikes and the plane is given a thorough inspection once it has landed.

man dennoch in einem Haus am besten vor Blitzeinschlägen geschützt.

Ein Blitzeinschlag in ein Flugzeug ist äußerst selten, durch gute Radargeräte an Bord umfliegen Piloten Gewitterwolken meist großräumig. Wenn es doch dazu kommt, wirkt die Metallhaut wie ein Blitzableiter. Da das Flugzeug in der Luft keine Erdung besitzt, tritt der Blitz ein, kurz danach aber auch wieder aus. Bei einem einzigen Blitzeinschlag bleibt es dabei aber meist nicht. Häufig wird ein Flugzeug in so einem Fall gleich von einer ganzen Serie, meist drei bis fünf, vereinzelt bis zu 25 Blitzen getroffen. Im Flugzeug selbst kann es zu einem dumpfen Knall kommen, Gefahr besteht aber keine – das Flugzeug ist ein „Faradayscher Käfig". Der Blitz schlägt ein, wird am Rumpf entlang geleitet und verlässt das Flugzeug wieder. Dennoch dokumentiert die Crew derartige Ereignisse, und, zurück am Boden, wird die Maschine durchgecheckt.

que les personnes sont les mieux protégées de la foudre.

Un impact de foudre sur un avion est relativement rare, car les pilotes disposent de radars embarqués performants qui leur permettent d'éviter les nuages orageux. Si la foudre frappe malgré tout un avion, son enveloppe métallique fait office de paratonnerre. Comme l'avion en vol n'est pas mis à la terre, la foudre y pénètre pour en ressortir aussitôt après. Un impact de foudre est rarement un phénomène isolé. Il est fréquent qu'un avion soit touché par toute une série d'éclairs, en général de trois à cinq, parfois de jusqu'à 25. Si l'on peut percevoir une explosion sourde à l'intérieur de l'avion, c'est sans danger car l'appareil tient lieu de « cage de Faraday » : la foudre le frappe, glisse le long du fuselage et s'éloigne. L'équipage consigne cependant l'incident et, de retour au sol, l'avion est inspecté.

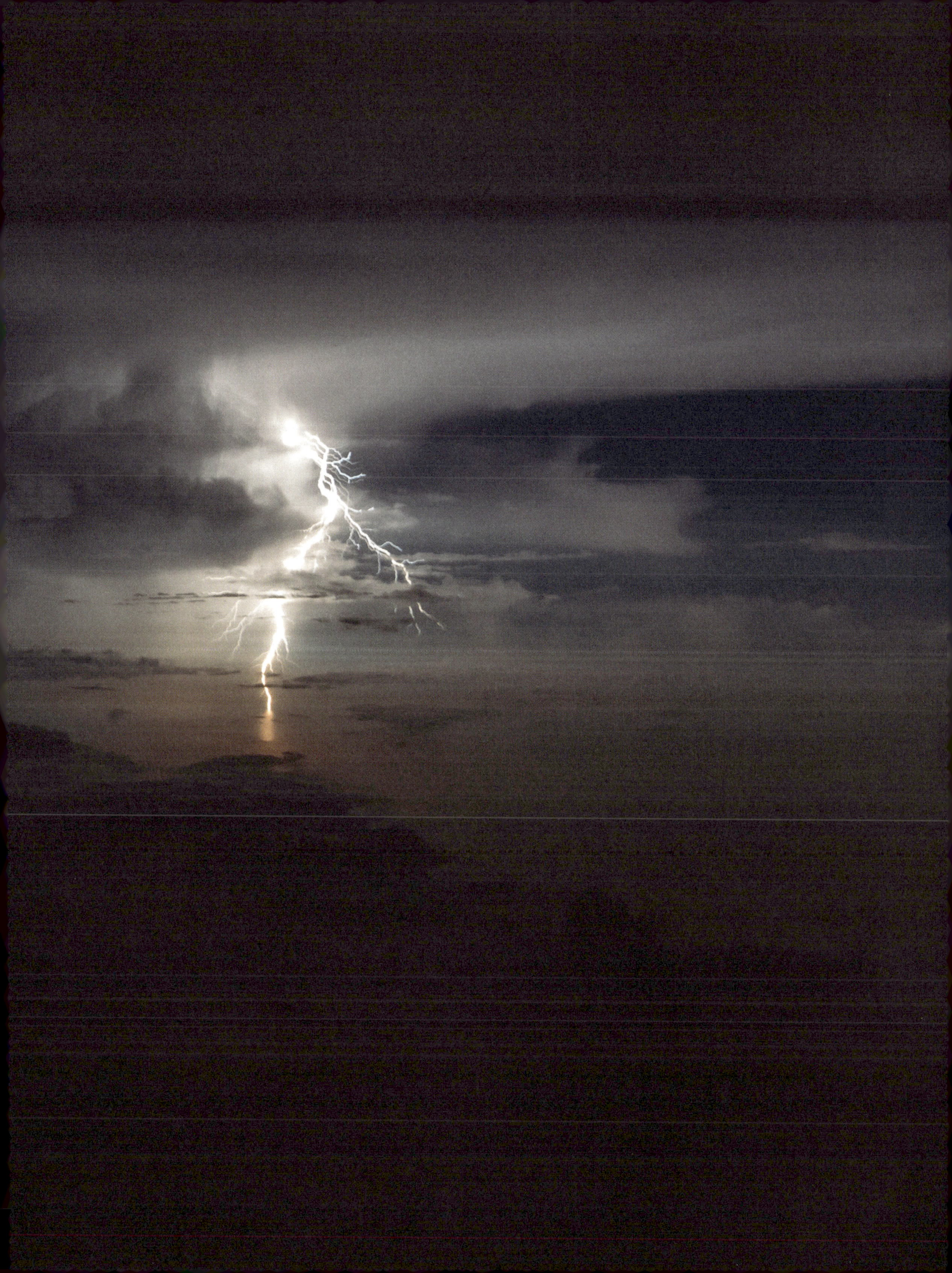

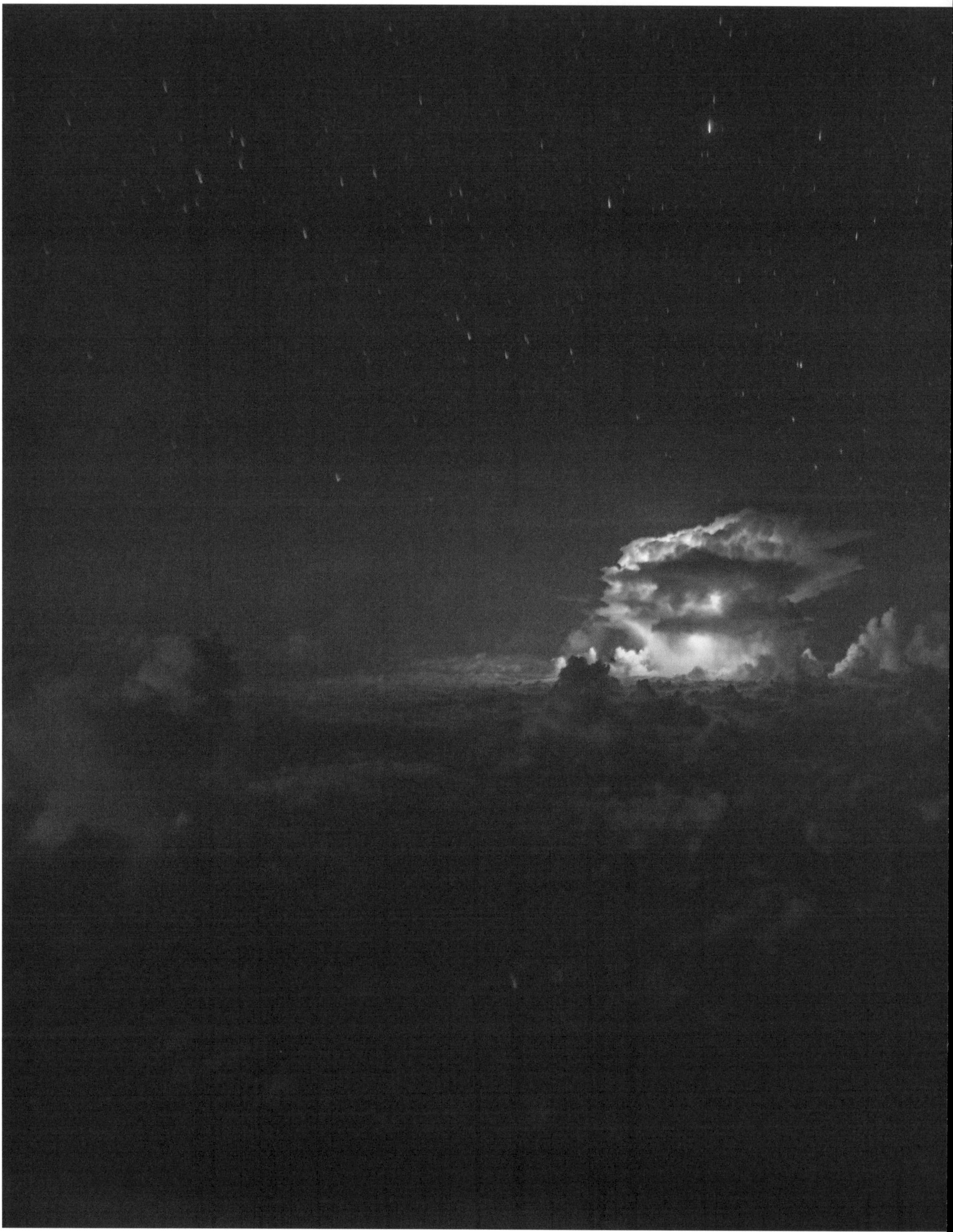

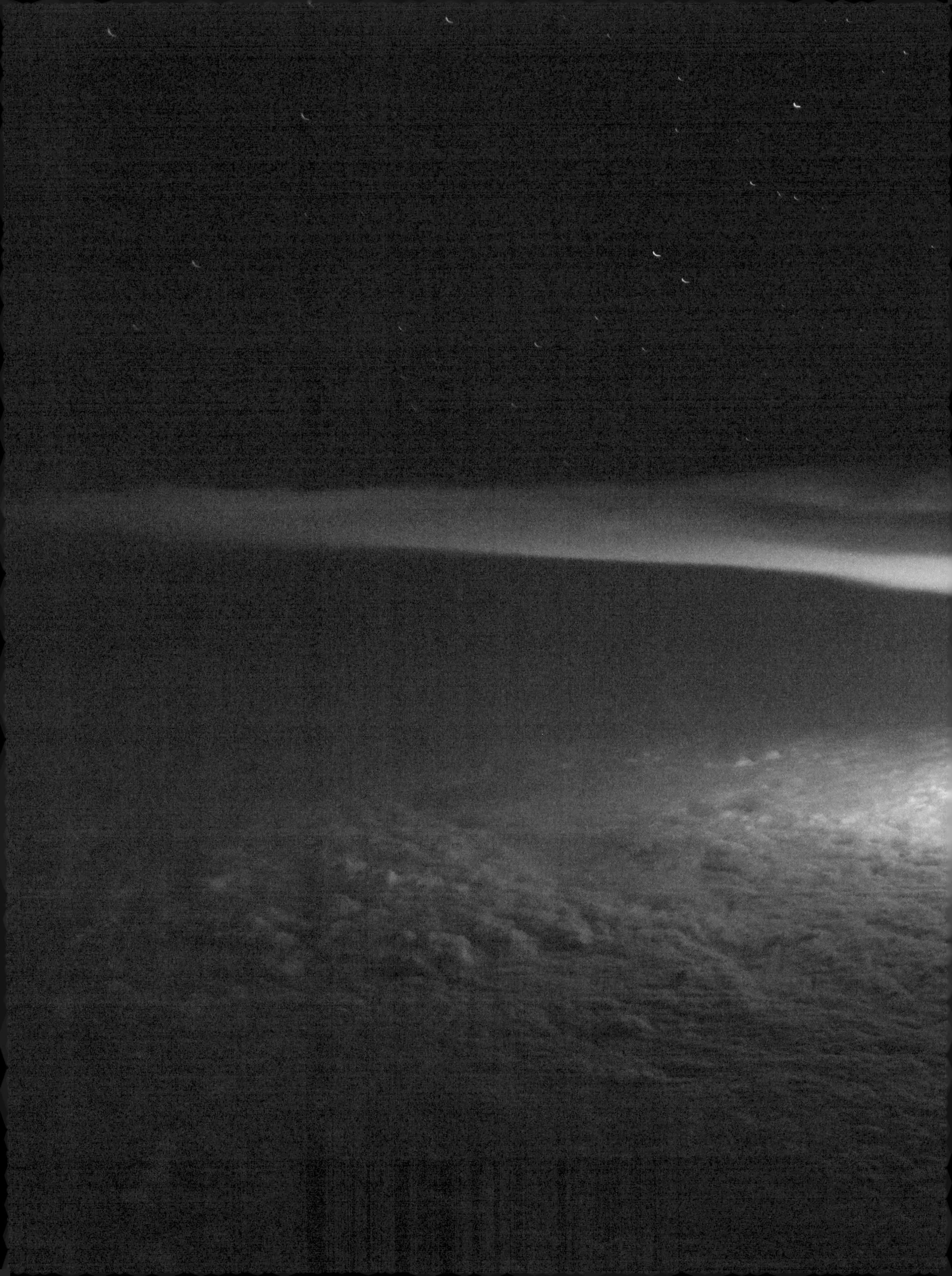

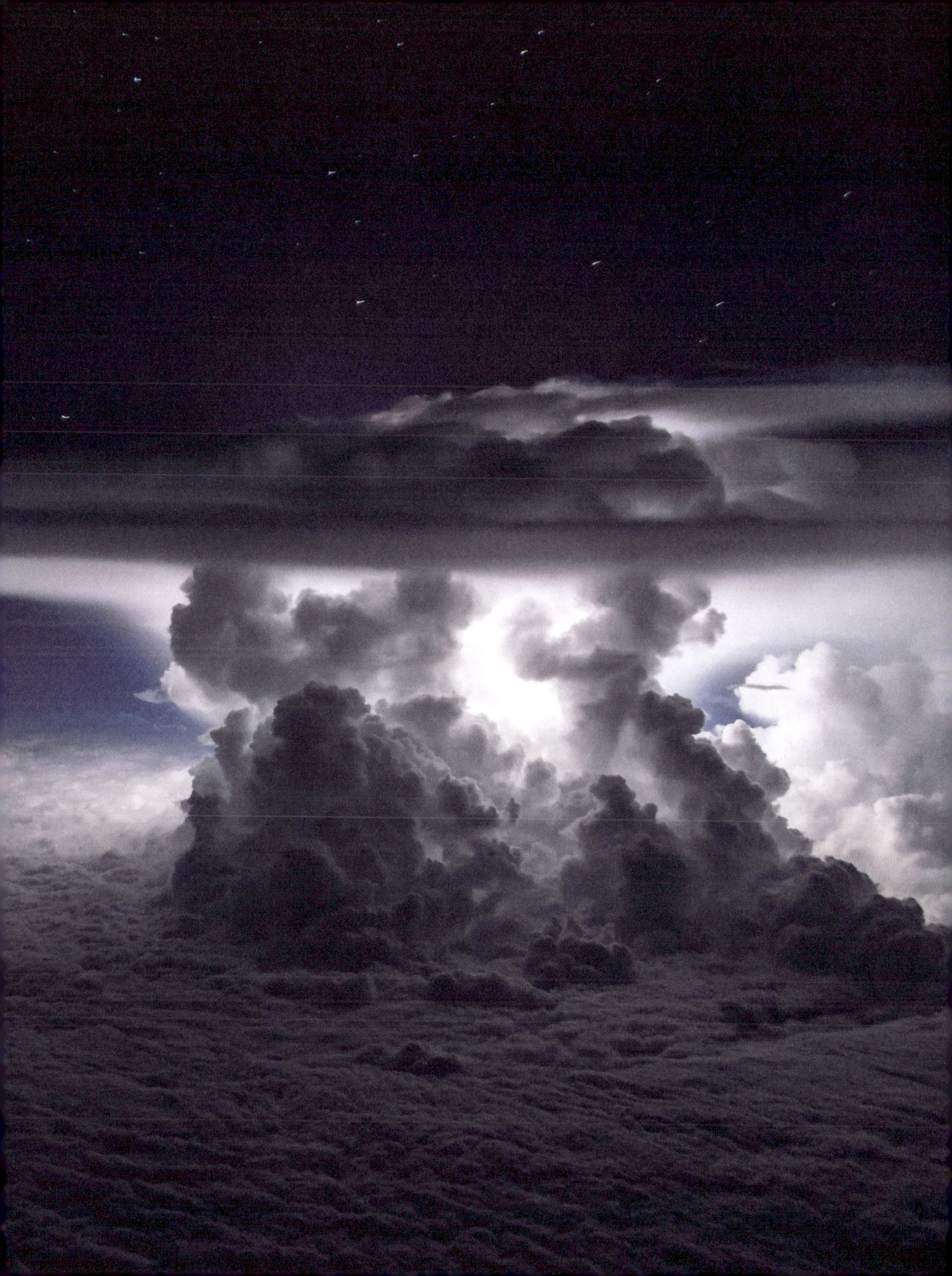

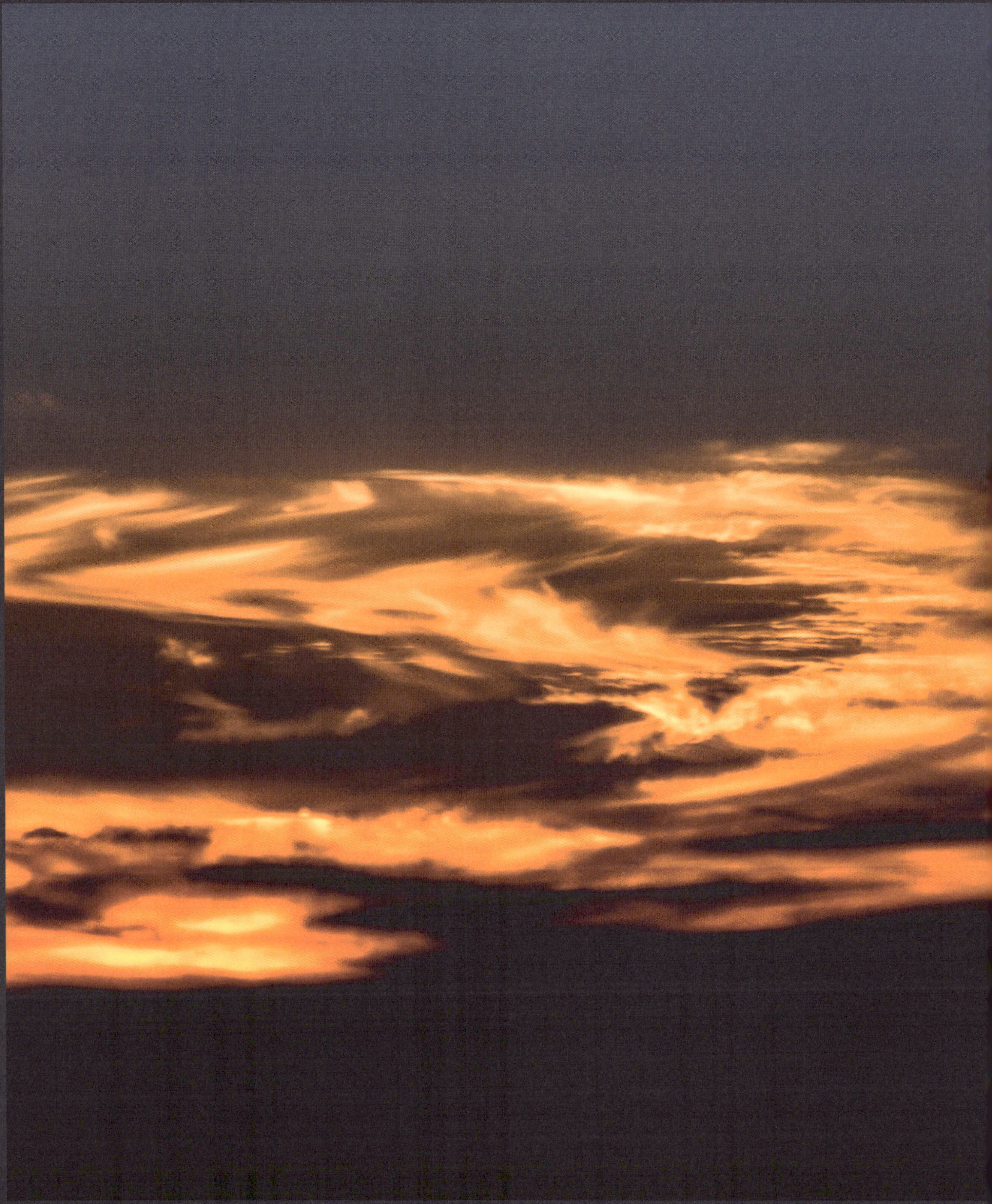

What is sheet lightning?

Sheet lightning is the term for flashes of light and brightness in the sky observed without accompanying thunder; it is sometimes referred to as heat lightning or summer lightning. The phenomenon occurs when storms are active, but at too great a distance for thunder to be audible; more often than not, no lightning bolts are visible neither. Instead, it can appear as if whole clouds are being lit by intermittent flashes of yellow or purple. As a general rule, sheet lightning can only be seen at night because daylight is too bright for lightning to be visible more than 60 miles away. So to observe this particular natural phenomenon, you need to be far away from urban lighting, have a far-reaching view, and be able to see thick storm clouds in an otherwise clear sky (without too much moonlight).

Was sind Wetterleuchten?

Wenn die Atmosphäre anfängt zu flackern und der Himmel in grellen Farben zuckt, es dabei aber mucksmäuschenstill bleibt – dann hat man es mit Wetterleuchten zu tun. Wetterleuchten entsteht durch Gewitter, die in weiter Entfernung aktiv sind – zu weit weg, um ein Donnern noch hören zu können. Meist sind nicht einmal Blitze zu erkennen. Eher hat man das Gefühl, als würden ganze Wolkenberge zuckend in Lila- und Gelbtönen aufleuchten und wieder erlöschen. Dabei kann man Wetterleuchten meist nur in der Nacht beobachten. Tagsüber ist es einfach zu hell, um Blitze in teils mehr als 100 Kilometern Entfernung sehen zu können. Weitere Voraussetzungen für ein derartiges Naturspektakel sind eine gute Fernsicht, hochreichende Gewitterwolken, ein klarer Himmel und kein störendes Stadt- oder Mondlicht.

Qu'est-ce qu'un « éclair de chaleur » ?

Quand l'atmosphère commence à scintiller et que le ciel se marbre de couleurs vives et tressaille, et ce sans le moindre bruit, on a affaire à des fulgurations ou éclairs de chaleur. Il s'agit en fait d'éclairs générés par des orages lointains, trop éloignés pour qu'on en perçoive les coups de tonnerre. En général, on ne voit même pas les éclairs, on a plutôt l'impression que des amas nuageux s'illuminent en clignotant dans des tons de mauve et de jaune avant de s'éteindre. Ce phénomène est visible presque uniquement de nuit. En journée, la trop forte luminosité empêche tout bonnement de voir des éclairs parfois distants de plus de 100 kilomètres. Pour profiter de ce spectacle de la Nature, il faut aussi une bonne visibilité, des nuages d'orage à forte extension verticale, un ciel dégagé et aucun éclairage urbain ni clair de lune parasites.

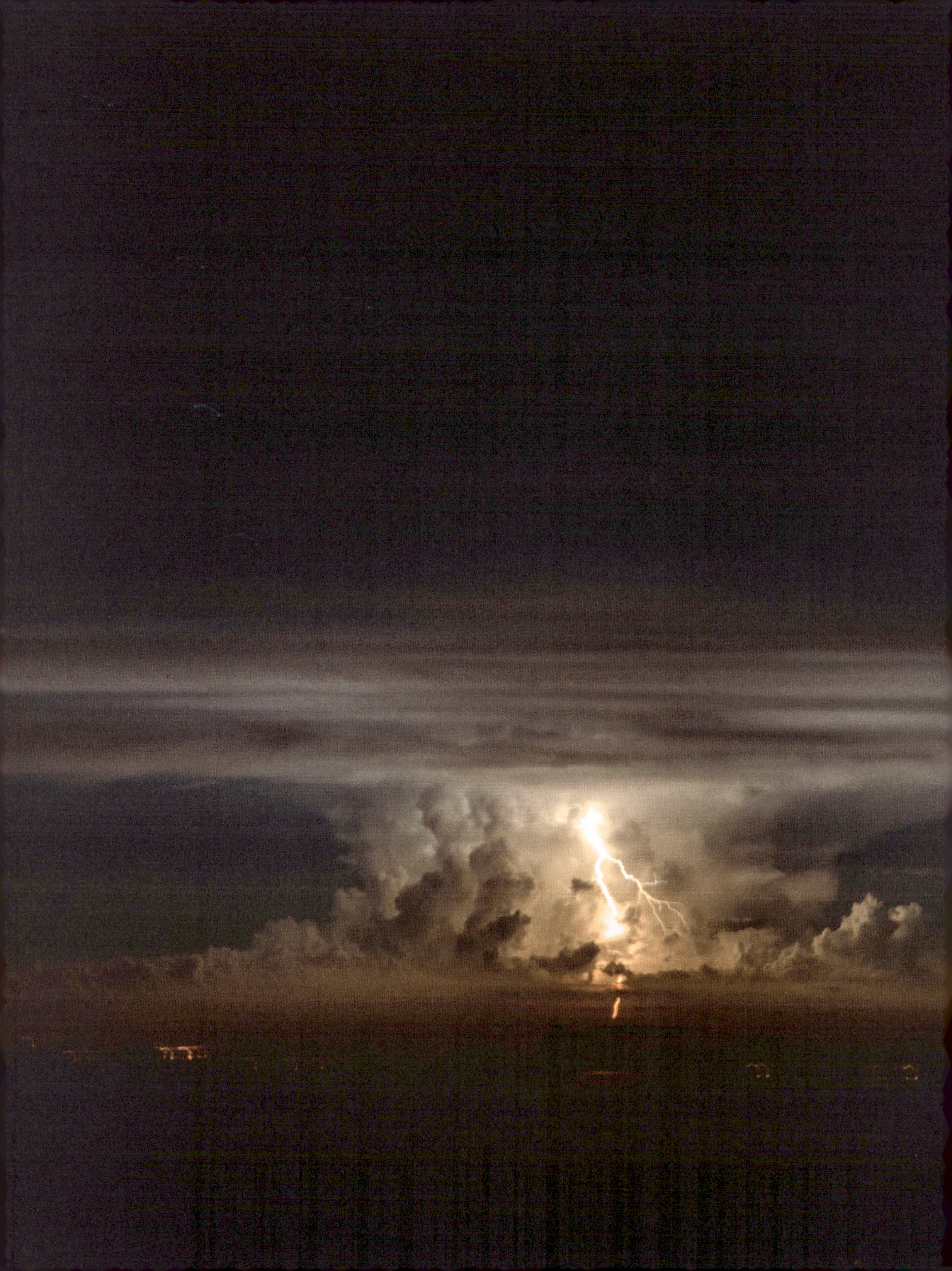

Big storms, but there's always a path through

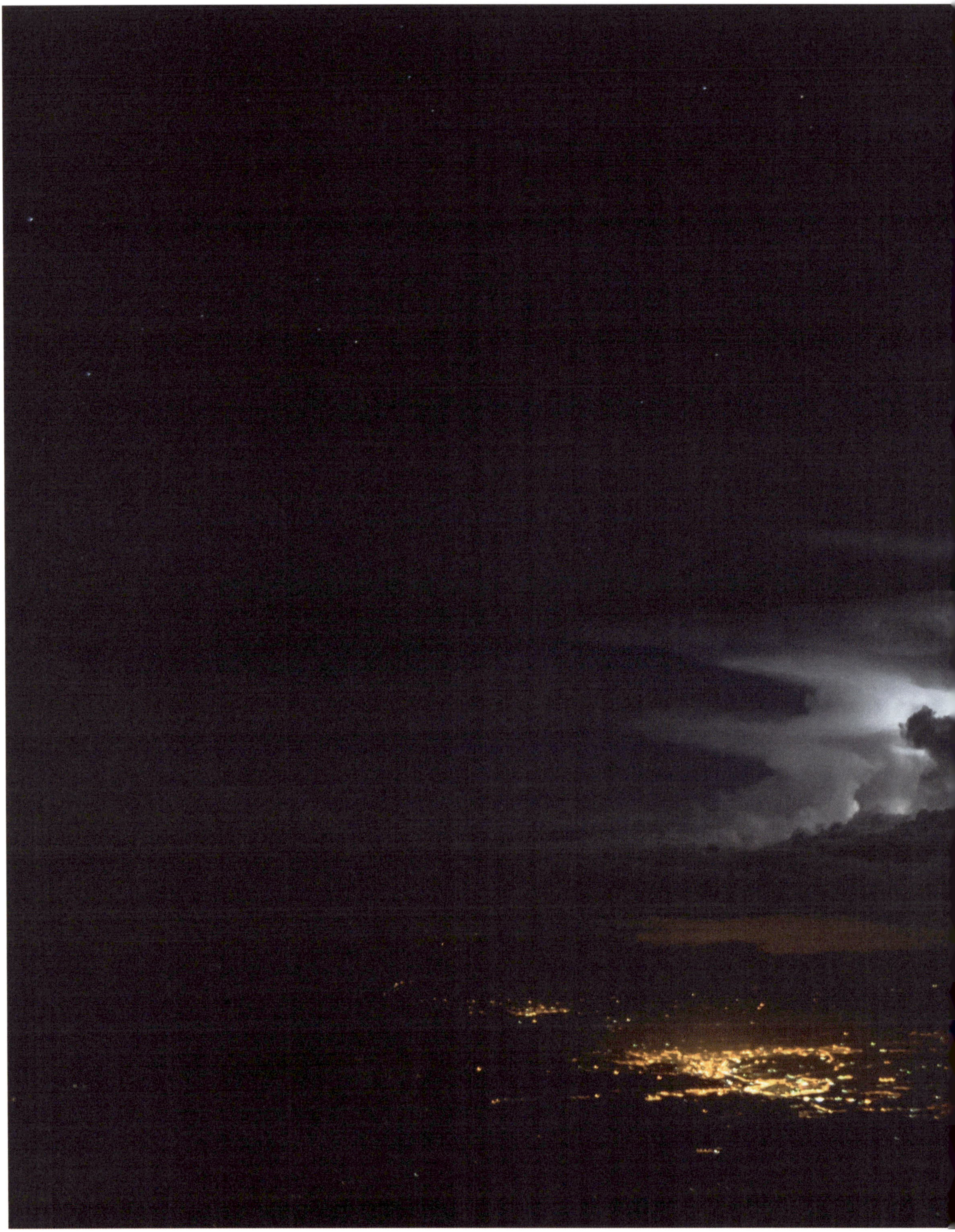

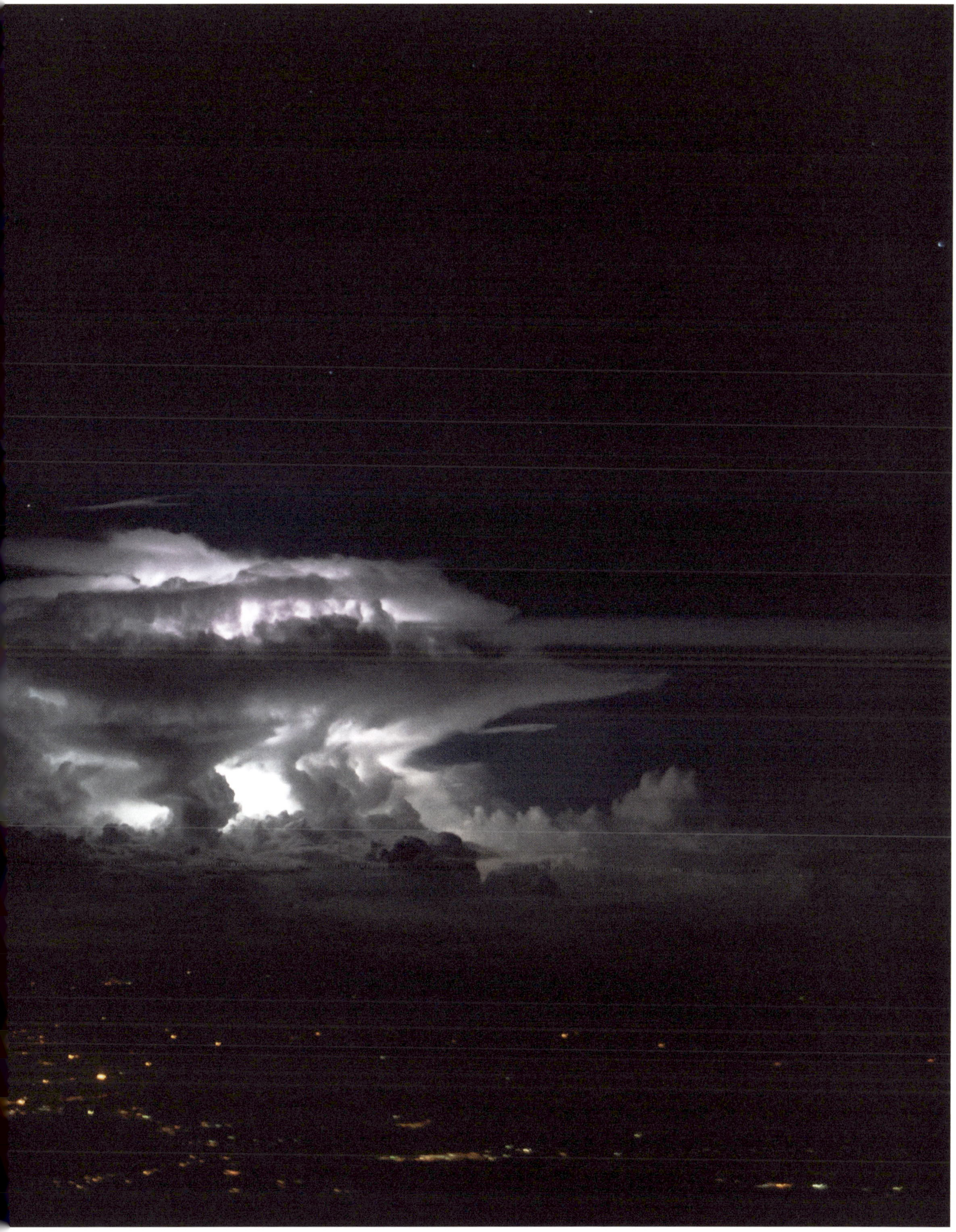

What are tornados?

Tornados are the fastest winds on earth, reaching maximum speeds of more than 300 mi/h. By way of comparison, most high-speed trains run at 185 mi/h. The strongest tornados on record occurred in May 1999 and 2013, both in Oklahoma, and both hitting peaks of 323 mi/h; with a diameter of 2,6 miles, the 2013 whirlwind was also the biggest ever sighted.

Generally small in scale, tornados are highly localised phenomena which are extremely difficult to predict, incredibly damaging to property and dangerous to life: again and again, these treacherous companions of supercell storms hoover up hapless victims. Rising up to 65,000 feet into the heavens and anywhere up to 100,000 feet wide, supercells are monstrous storm clouds which can produce tornados if a range of factors combine:

extremely high windspeeds inside the storm cloud

great differences in temperature between the top and bottom of the cloud

high humidity in the supercell

Was sind Tornados?

Tornados sind die schnellsten Winde der Welt. Es können dabei Windspitzen von über 500 km/h auftreten. Zum Vergleich: Die meisten Schnellzüge kommen auf etwa 300 km/h. Die bisher stärksten Tornados wurden jeweils im Mai 1999 und 2013 in Oklahoma mit Windspitzen von circa 520 km/h gesichtet. Mit einem Durchmesser von 4,2 Kilometern war derselbe Tornado 2013 auch der größte je beobachtete.

Tornados sind kleinräumige, aber tückische Winde, denn sie sind sehr schwer vorherzusagen, richten schlimme Schäden an und fordern immer wieder Todesopfer. Sie entstehen unter gewaltigen Gewitterwolken, den sogenannten Superzellen. Diese Monstergewitter reichen bis zu 20 Kilometer in die Höhe und können 30 Kilometer breit sein. Verschiedene Faktoren müssen zusammenkommen, damit sich Tornados, natürliche Riesenstaubsauger, am unteren Rand der Gewitterwolke bilden können:

extrem hohe Windgeschwindigkeiten in der Gewitterwolke

große Temperaturunterschiede zwischen oben und unten

sehr viel Feuchtigkeit in der Wolke

Qu'est-ce qu'une tornade ?

Les tornades sont les vents les plus rapides de la planète, avec des rafales pouvant dépasser les 500 km/h. À titre de comparaison, la plupart des trains à grande vitesse roulent à environ 300 km/h. Les tornades les plus violentes enregistrées à ce jour, ont été observées en Oklahoma en mai 1999 et 2013, avec des rafales d'environ 520 km/h. D'un diamètre de 4,2 kilomètres, celle de mai 2013 est aussi la plus étendue jamais observée.

Les tornades sont de petites dimensions mais traîtres car elles sont difficilement prévisibles, provoquent de lourds dégâts et font régulièrement des victimes. Elles prennent naissance sous d'énormes nuages orageux donnant des orages dits supercellulaires. Ces perturbations monstres atteignent jusqu'à 20 kilomètres de hauteur et 30 kilomètres de largeur. Différents facteurs doivent être réunis pour que les tornades, ces aspirateurs naturels géants, puissent se former :

des vitesses de vent extrêmes au sein du nuage orageux

des écarts de températures importants entre le haut et le bas du nuage

une forte humidité dans le nuage

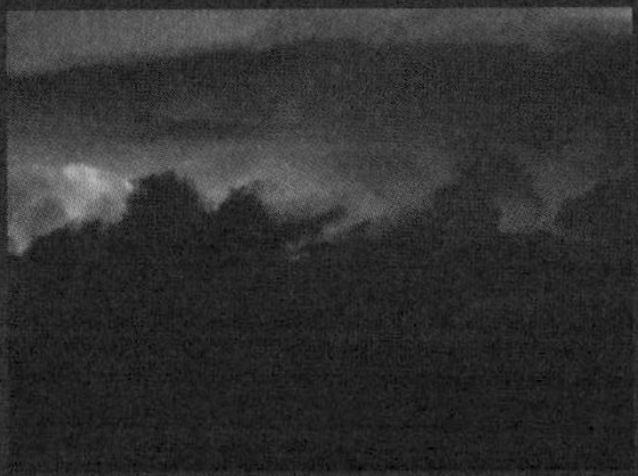

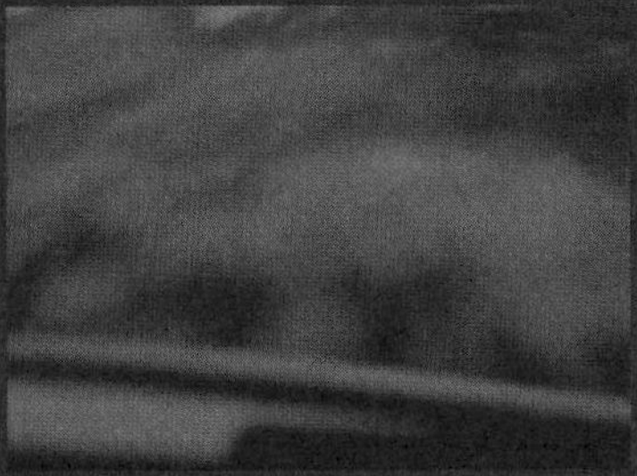

Another element in the emergence of a tornado is that the winds in the lower section of the cloud are blowing in a different direction to those at the top. When this occurs, meteorologists talk of a wind shear, a highly volatile vortex which is usually accompanied by torrential rainfall and heavy hail showers.

As long as tornados don't make contact with the surface of the earth, the vortex is visible in the sky, but doesn't pose any particular danger. Once the funnel hits the ground, though, it develops into a brutal, destructive maelstrom which sucks in everything in the immediate vicinity. Incidentally, tornados are not exclusive to the USA and can occur all over the world, both on land and over water (where they are called waterspouts). Nevertheless, the extreme temperature differentials which can occur over the north American landmass and the fact that large swathes of US territory are not separated from each other by mountain ranges means that series of strong tornados are often to be seen there, especially in Tornado Alley in the Midwest.

Außerdem weht bei der Entstehung eines Tornados der Wind im unteren Teil der Wolke in eine andere Richtung als weiter oben, die sogenannte Windscherung tritt auf – es entstehen brandgefährliche Luftwirbel, meist zusammen mit heftigem Platzregen und schwerem Hagelschlag.

Wenn Tornados den Erdboden nicht berühren, sind sie ungefährlich, sehen nur etwas seltsam aus – man nennt sie Funnel. Sobald der Rüssel allerdings Bodenkontakt bekommt, wird er wütend und zerstörerisch, ein Sog hat sich gebildet. Tornados treten übrigens nicht nur in den USA auf, es gibt sie auf der ganzen Welt, auf dem Land und auf Wasser – man nennt sie dann Wasserhosen. Weil in Amerika zum Teil extreme Temperaturunterschiede auftreten können und große Teile des Landes nicht durch Gebirgszüge getrennt sind, werden dort allerdings gehäuft stärkere Tornados gesichtet, vor allem in der sogenannten „Tornado Alley" im Mittleren Westen.

De plus, une tornade se développe en association avec un cisaillement de vent – qui souffle dans une direction dans le bas du nuage, et dans une autre dans le haut du nuage. Il en résulte des tourbillons d'air dangereux au plus haut point qui s'accompagnent en général d'averses violentes et de fortes précipitations de grêle.

Tant qu'une tornade ne touche pas le sol, elle est inoffensive, son apparence est juste un peu étrange. On l'appelle « tuba ». Mais dès que son embouchure touche le sol, il se crée une aspiration, et la tornade se déchaîne, détruisant tout sur son passage. Les tornades sévissent dans le monde entier, sur les terres comme à la surface des océans – dans ce dernier cas, on les appelle trombes marines. Aux États-Unis cependant, comme des écarts de température extrêmes peuvent survenir et qu'une grande partie du territoire n'est pas cloisonnée par des chaînes montagneuses, on observe souvent dans ce pays des tornades relativement puissantes, surtout dans la « Tornado Alley », dans le Midwest.

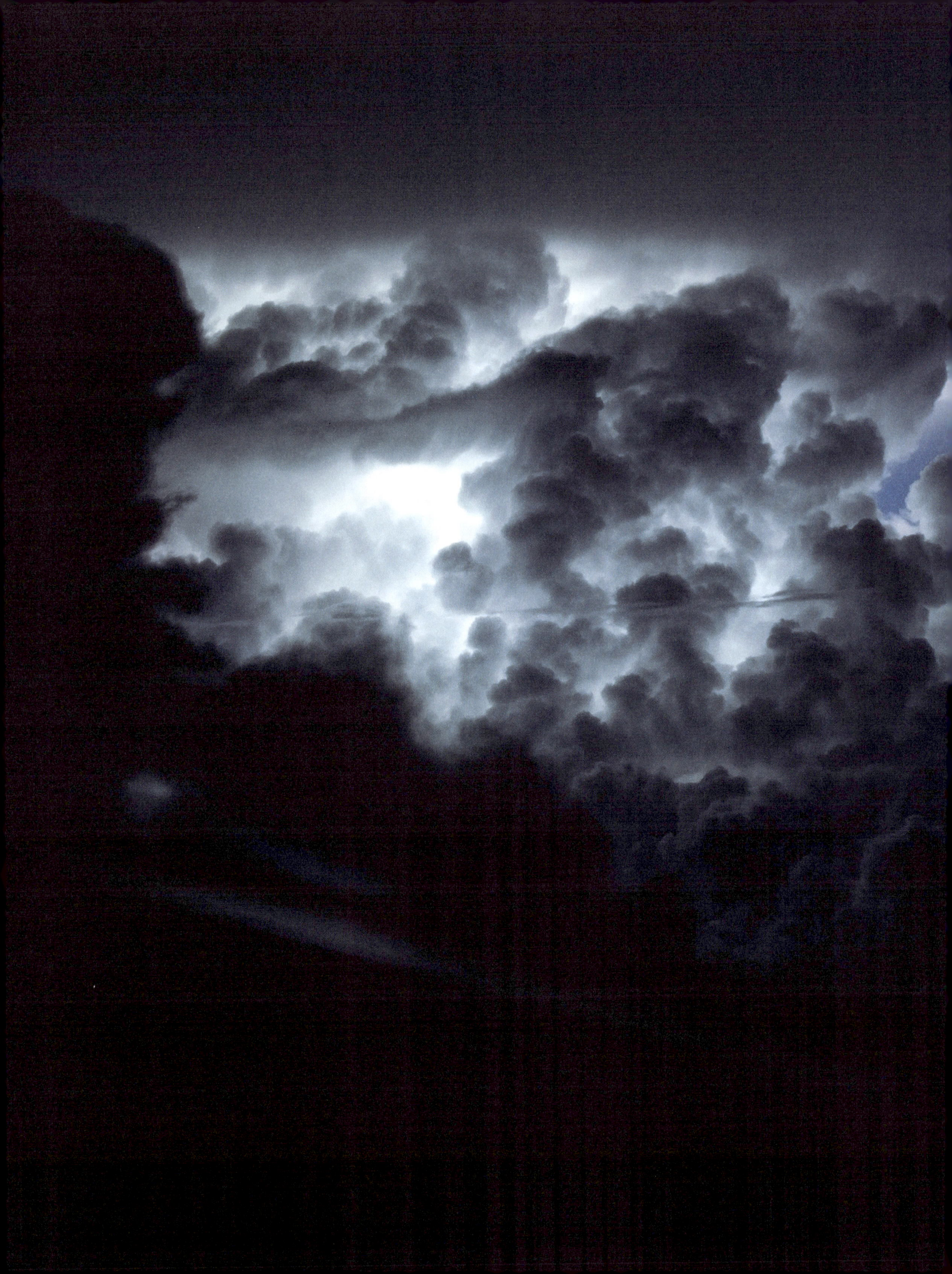

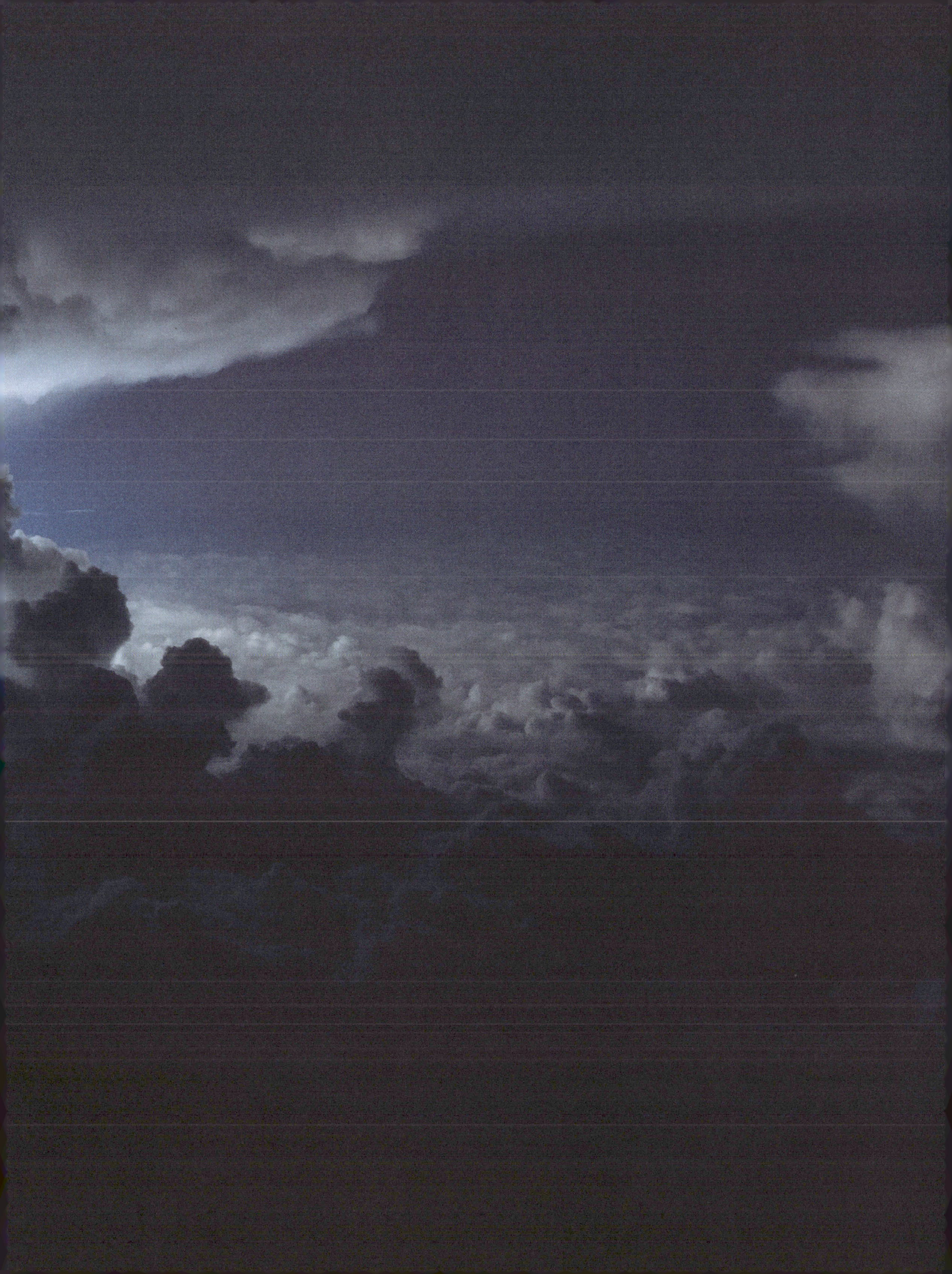

How do hurricanes happen?

The word hurricane is used to refer to tropical cyclones occuring over the Atlantic Ocean between May and September. Although they don't reach wind speeds comparable to those of tornados, hurricanes are gigantic in scale and have the potential to cause widespread damage. In 2011, Hurricane Irene was the size of Europe and when a storm like this—with peak wind speeds of 190 mi/h—makes landfall, the effect is extremely destructive: torrential rainfall and storm tides are the result; the danger to human life is significant.

Hurricanes occur over warm seas once the water temperature has reached a minimum of 81°F (27°C): sustained exposure to the sun causes large amounts of water to evaporate and form clouds. The low pressure this causes above the water's surface sucks in air sideways and the currents streaming inwards begin to whirl—much like the way water forms a vortex around a plughole. This leads to a gigantic cyclone whose circular centre, the eye, is almost windless.

Photographed from satellites in orbit, hurricanes have an eminently recognisable form; they can occur with varying frequency over most of the world's oceans

Wie entstehen Hurrikans?

Hurrikans sind tropische Wirbelstürme, die auf dem Atlantik zwischen Mai und September entstehen. Hier werden zwar nicht so hohe Windgeschwindigkeiten wie bei Tornados erreicht, aber sie sind riesig und bringen große Zerstörung. Hurrikan „Irene" von 2011 hatte die Größe von ganz Europa. Wenn solch ein Sturm mit Windspitzen von 300 km/h die Küste erreicht, ist er äußerst zerstörerisch und kann Menschenleben fordern, er bringt gewaltige Regenmengen und Flutwellen an den Küsten.

Hurrikans bilden sich über dem warmen Meer bei einer Wassertemperatur von mindestens 27 Grad Celsius. Durch hohe Sonneneinstrahlung verdunstet dabei viel Wasser, steigt auf und bildet Wolken. Durch den so entstandenen Unterdruck über der Wasseroberfläche wird Luft von der Seite nachgesaugt. Diese nachströmende Luft beginnt sich zu drehen wie ein Strudel, wenn man den Stöpsel aus der Badewanne zieht. Ein riesiger Wirbel entsteht, dessen Drehkreuz – das Auge des Sturms – aber fast völlig windstill ist.

Hurrikans haben – auf Satellitenbildern betrachtet – eine sehr charakteristische Form und tauchen in unterschiedlicher

Comment les ouragans se forment-ils ?

Les ouragans sont des phénomènes tourbillonnaires tropicaux, qui se forment sur l'océan Atlantique entre mai et septembre. Ils n'atteignent pas les mêmes vitesses que les tornades, mais sont gigantesques et provoquent des dégâts considérables. L'ouragan « Irene » de 2011 avait la taille de l'Europe. Quand pareille tempête atteint le littoral avec des rafales de 300 km/h, les pluies diluviennes et les raz-de-marée qui s'abattent sur les côtes peuvent faire des morts.

Les ouragans se forment au-dessus de la mer quand on enregistre au moins 27 °C dans ses eaux de surface. Le fort ensoleillement entraîne une évaporation importante, qui se transforme en nuages. La dépression résultant à la surface de l'eau entraîne une aspiration latérale de l'air, qui se met alors à s'enrouler en spirale, comme le tourbillon visible quand on retire un bouchon de baignoire. Dans le centre de rotation de ce gigantesque tourbillon, dans « l'œil du cyclone », il règne un calme presque plat.

Les ouragans ont une forme très caractéristique, comme on peut le voir sur les images satellite. Ils surviennent plus ou mois fréquemment au-dessus

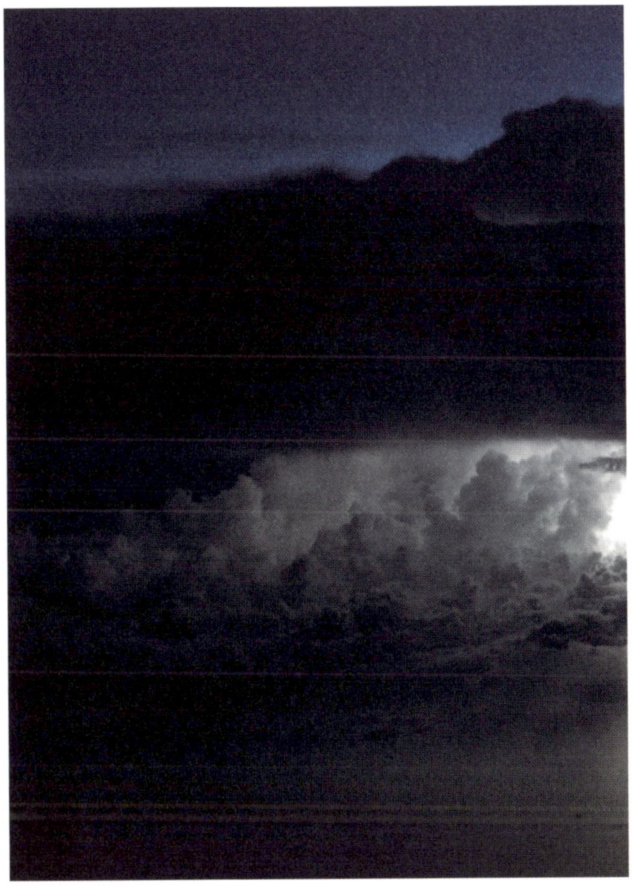

which reach the necessary temperatures. If they form in east Asia, they are referred to as typhoons and are simply called tropical cyclones in the Gulf of Bengal. Not infrequently, hurricanes travel as far as Europe, as did Ophelia in 2017. Tropical-like cyclones ("Medicanes") can also occur in the Mediterranean basin, but are far weaker than hurricanes.

Häufigkeit und mit unterschiedlichen Bezeichnungen auf allen Weltmeeren mit ausreichender Wassertemperatur auf. In Ostasien werden sie Taifune genannt, im Golf von Bengalen Zyklone. Manchmal verirren sich Hurrikans sogar nach Europa, wie „Ophelia" 2017. Auch der Mittelmeerraum kennt mit dem sogenannten Medicane ein ähnliches, jedoch weitaus schwächer ausgeprägtes Wetterphänomen.

de toutes les mers du globe suffisamment chaudes en surface et portent des noms différents selon les régions. En Asie de l'Est, on parle de typhons, dans le golfe du Bengale de cyclones. Il arrive parfois que des ouragans s'égarent jusqu'en Europe, comme « Ophelia » en 2017. La Méditerranée connaît un phénomène météorologique semblable, mais de bien plus faible ampleur, le « médicane » (contraction de Méditerranée et de *hurricane*, ouragan en anglais).

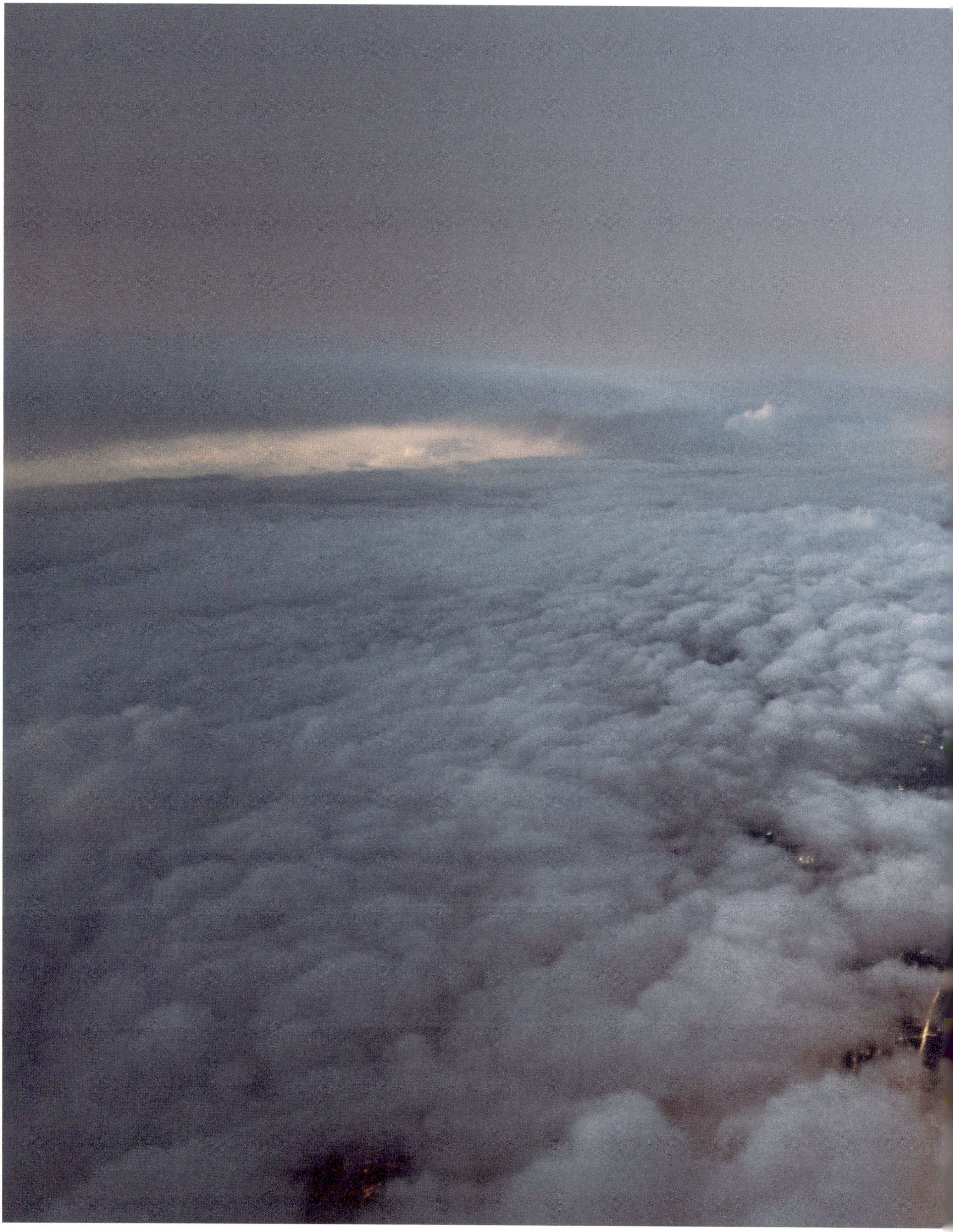

How far up do planes fly?

Looking out of the window into the clouds and watching how these intangible aerosols change, how water turns into ice as the plane climbs higher and higher… you don't need to be a meteorologist to be fascinated by that! What's interesting though, is that, beyond a certain altitude, it soon becomes very difficult to estimate how far up you are. You can't make out people, cars look like ants and whole cities might just as well be made of Lego: but how high up are you? What is the typical cruising altitude for an airliner?

The length of the flight has an effect: short-haul routes are generally flown at around 26,000 feet; on long-haul flights, passengers get somewhat closer to space, with typical cruising altitudes of 40,000 feet. In general, at this height, the plane is over the top of the clouds, with only a few cirrus made entirely of ice crystals erring up this far (and, at most, a little higher). This type of cloud never contains any rain, but can create some turbulence.

The only airliner to fly at higher altitudes was Concorde: the first supersonic passenger jet, which last flew in October 2003, cruised at just under 65,000 feet up at a maximum speed of

In welcher Höhe fliegen Flugzeuge?

Am Fenster sitzen, Wolken beobachten, herausbekommen, wann aus Wasserwolken Mischwolken und später Eiswolken werden – spannend, und das nicht nur für Experten. Beeindruckend aber auch, dass sich ab einer bestimmten Höhe nur noch sehr schlecht schätzen lässt, wie hoch das Flugzeug eigentlich fliegt. Menschen lassen sich gar nicht mehr erkennen, Autos sehen aus wie Ameisen und Orte wie aus Legosteinen erbaut. Aber in welcher Höhe fliegen Passagierflugzeuge eigentlich?

Je nach Länge des Flugs gibt es Unterschiede: Kurze Strecken werden meist in einer Höhe von 8000 Metern geflogen. Bei Langstrecken kommt man dem Weltraum etwas näher, meist fliegen die Piloten in etwa 12.500 Metern Höhe. Damit befindet man sich meistens schon über den Wolken. Nur Cirren, Wolken also, die ausschließlich aus Eiskristallen bestehen, sind auf dieser Höhe oder höher zu finden. Diese Wolken bringen übrigens nie Regen, können allerdings zu Turbulenzen führen.

Die Concorde, das erste Überschall-Passagierflugzeug, die ihren letzten Flug im Oktober 2003 startete, war das einzige Flugzeug, das in einer Höhe von

À quelle altitude les avions volent-ils ?

Assis côté hublot, on observe les nuages et l'on voit les nuages de gouttelettes d'eau devenir des nuages mixtes puis des nuages de cristaux de glace. Voilà qui est passionnant, et pas uniquement pour les spécialistes. Il est aussi fascinant de constater qu'à partir d'une certaine distance du sol, on a bien du mal à évaluer l'altitude de vol. On ne voit plus les personnes, les véhicules ont la taille de fourmis, et les zones urbanisées ressemblent à des villes en Lego. Mais à quelle altitude les avions de ligne volent-ils au juste ?

L'altitude de vol est fonction de la distance à parcourir : elle est en général de 8 000 mètres pour les vols court-courriers, et de 12 500 mètres pour les vols long-courriers. L'on se trouve alors normalement au-dessus de la couverture nuageuse. À cette altitude, les seuls nuages que l'on puisse rencontrer sont des cirrus. Entièrement constitués de cristaux de glace, ils n'engendrent jamais de pluie mais peuvent être à l'origine de turbulences.

Premier avion de ligne supersonique, le Concorde, qui a effectué sa dernière mission en octobre 2003, était le seul avion volant à près de 20 kilomètres d'altitude. Il atteignait une

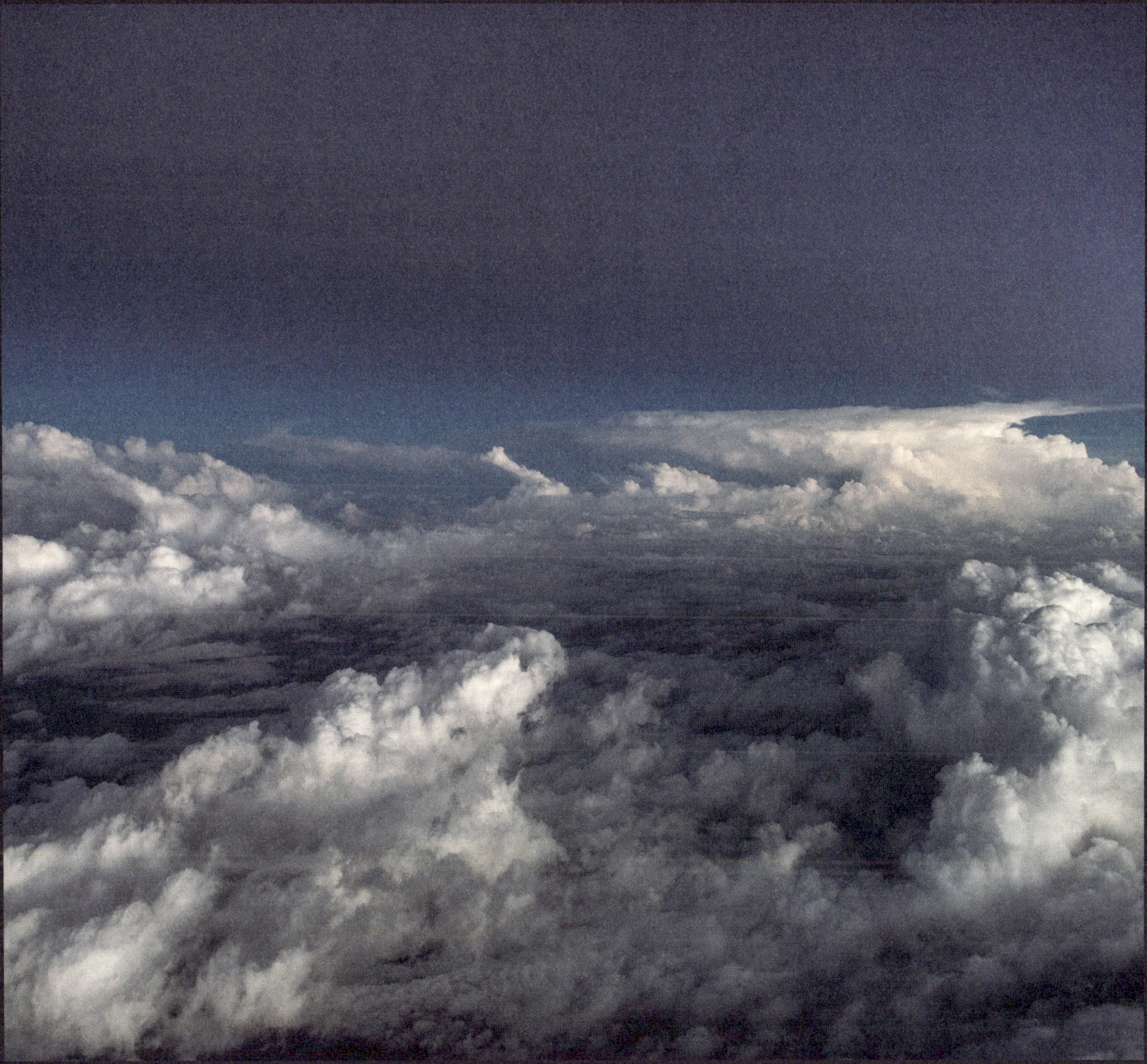

1333 mi/h (that's the same speed as a bullet). Able to travel from Paris to New York in three or three and a half hours at most, it was retired following a crash in 2000 and due to its high running costs.

knapp 20 Kilometern flog. Sie erreichte Höchstgeschwindigkeiten von 2146 km/h und war damit so schnell wie eine Gewehrkugel. Die Flugzeit zwischen Paris und New York betrug gerademal 3 bis 3,5 Stunden. Durch den Absturz einer Maschine im Juli 2000 und die hohen Kosten wurde die Concorde allerdings mittlerweile eingestellt.

vitesse de croisière maximale de 2 146 km/h, pratiquement celle d'une balle de fusil. La durée de vol entre Paris et New York était tout juste de 3 à 3,5 heures. À la suite d'un crash en juillet 2000, cet avion par ailleurs cher à entretenir a été retiré de la circulation.

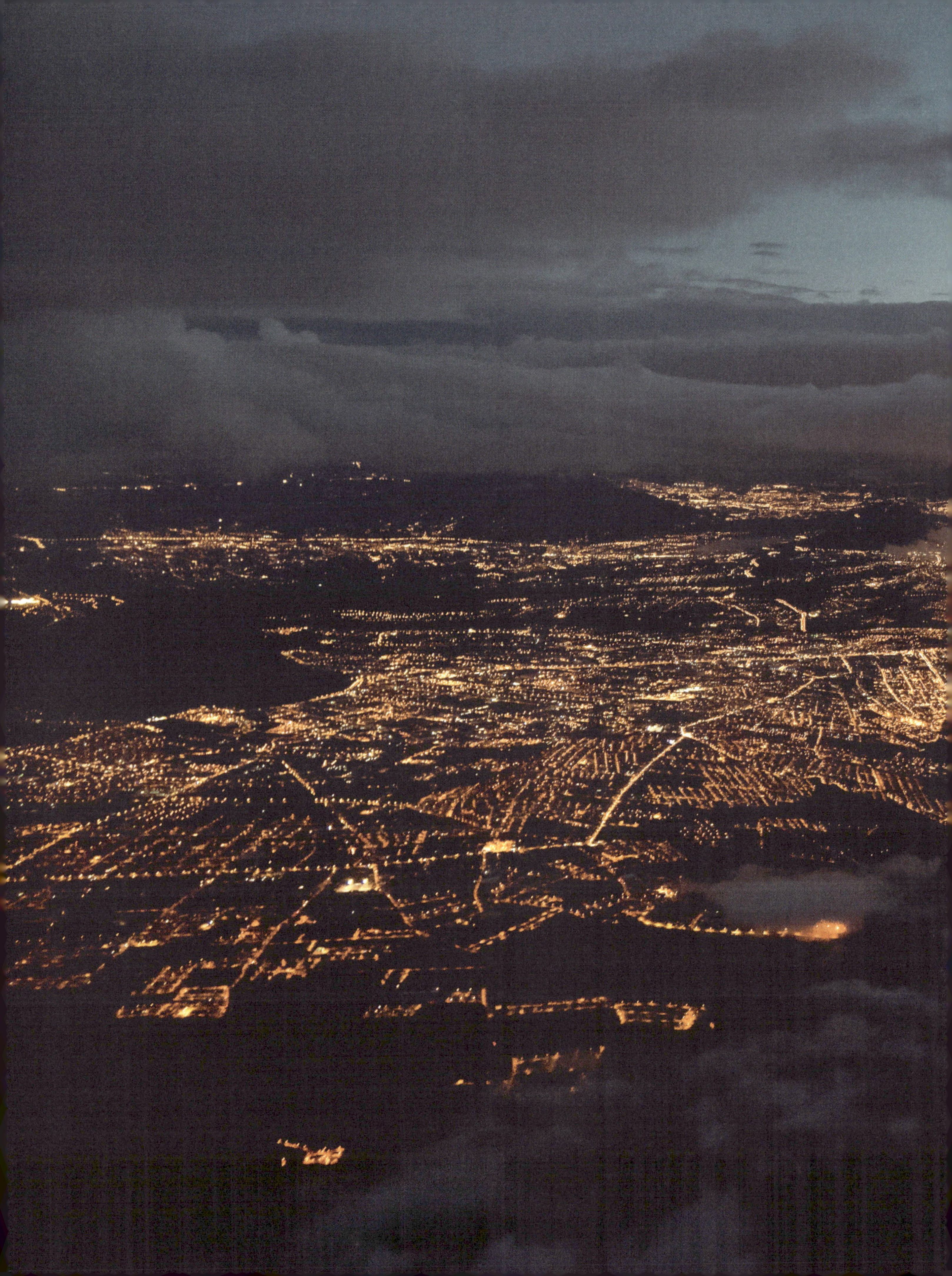

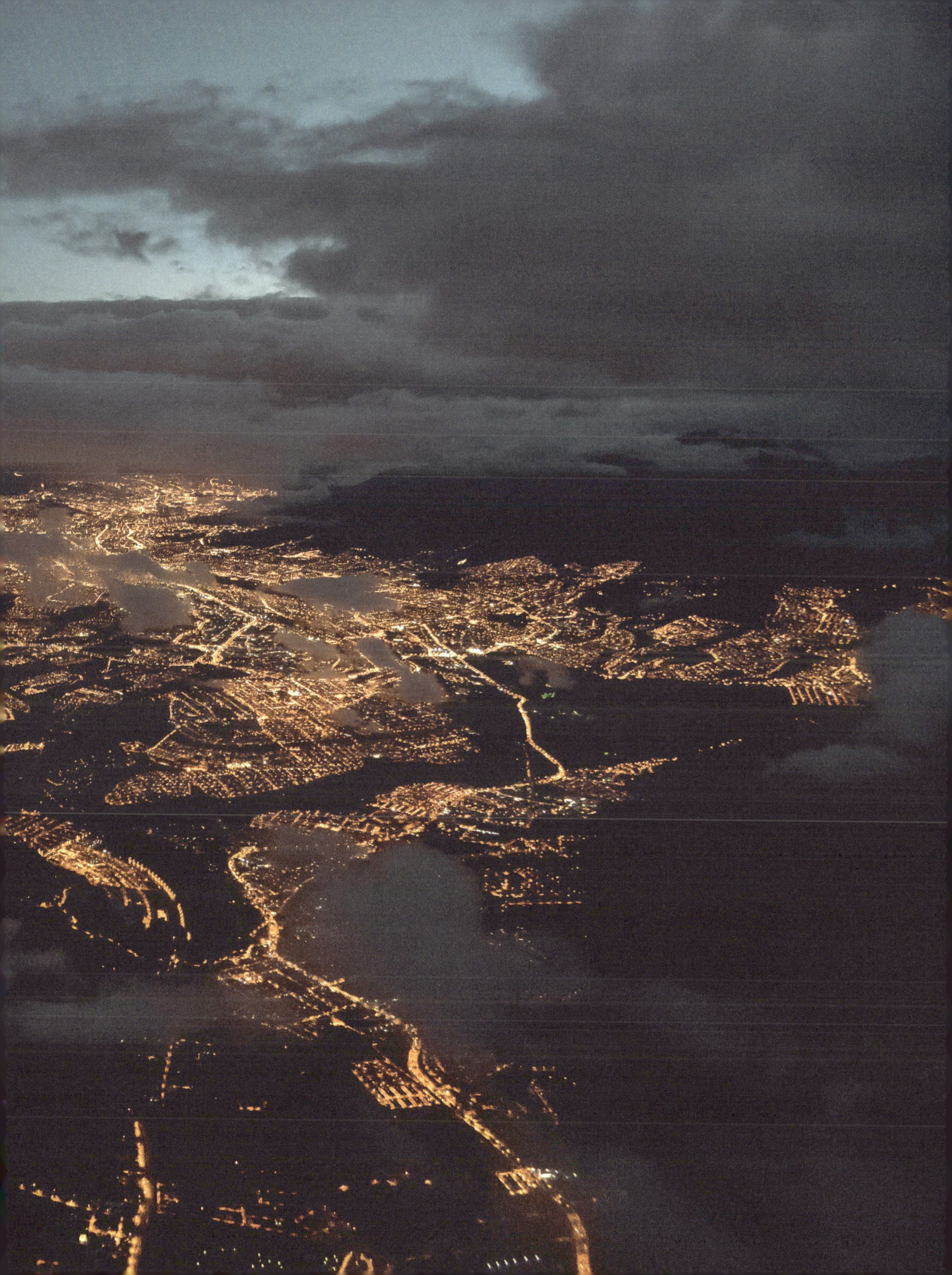

Why does the sky turn red at dawn and dusk?

Sunlight takes on different colours at different times of day: while it generally appears to be bright yellow or hot white around midday, in the mornings and evenings, warmer reddish tones can be dominant. Yet if the sun itself doesn't change its colour and always shines at the earth with the same power, why does its light look different according to the time of day?

The colours of sunlight are those of the rainbow: red, orange, yellow, green, blue, and violet. Each one occupies different wavelengths on the electromagnetic spectrum and that is what allows our eyes to recognise them as distinct colours. Yet when sunlight hits the earth's atmosphere, it comes across large amounts of particles—generally dust or water molecules. These particles reflect or refract the light of the sun depending on their own composition and the angle at which sunrays hit them. This means that the colours in the light with their various wavelengths are dispersed to varying degrees.

At midday, sunlight approaches our atmosphere vertical, reducing the distance it travels to reach the surface of the earth; in the morning and evening, light

Wie entstehen Morgen- und Abendrot?

Das Licht der Sonne nimmt manchmal zu unterschiedlichen Tageszeiten verschiedene Farben an. Mittags wird es meist als grellgelb oder weiß wahrgenommen, wohingegen morgens und abends eher wärmere, rötliche Farben dominieren können. Wie aber kommt es dazu, wo doch die Sonne ihre Farbe gar nicht verändert und immer mit der gleichen Strahlkraft leuchtet?

Sonnenlicht besteht aus den Regenbogenfarben Rot, Orange, Gelb, Grün, Blau und Violett. All diese Farben haben im elektromagnetischen Spektrum unterschiedliche Wellenlängen – weswegen unser Auge sie auch als Farben erkennen kann. Tritt nun das Sonnenlicht in unsere Erdatmosphäre ein, trifft es dort auf eine große Menge winzig kleiner Teilchen wie Staub oder Wassertröpfchen. An diesen Teilchen werden die Sonnenstrahlen unterschiedlich reflektiert oder gebrochen – je nach Beschaffenheit des Teilchens oder dem Einstrahlungswinkel des Lichts. Dabei werden auch die unterschiedlichen Farbanteile mit ihren unterschiedlichen Wellenlängen verschieden stark abgelenkt.

Mittags tritt das Sonnenlicht senkrecht von oben in unsere

Comment l'aube et le crépuscule se forment-ils ?

La couleur de la lumière solaire varie selon le moment de la journée. À midi, on la perçoit comme jaune vif ou blanche, alors que le matin et le soir, des tons plus chauds de rouge peuvent dominer. Comment est-ce possible ? En effet, ni la couleur du soleil ni l'intensité de son rayonnement ne changent.

La lumière solaire est constituée des couleurs de l'arc-en-ciel rouge, orange, jaune, vert, bleu et violet. Toutes ces couleurs ont des longueurs d'ondes différentes dans le spectre électromagnétique, ce qui permet à notre œil de les percevoir en tant que couleurs. Quand la lumière solaire pénètre dans l'atmosphère terrestre, elle rencontre une quantité prodigieuse de micropoussières et de gouttelettes d'eau microscopiques. La manière dont ces particules réfléchissent ou réfractent la lumière varie en fonction de la nature des particules et de l'angle d'incidence de la lumière. N'ayant pas la même longueur d'onde, les différentes couleurs sont elles-mêmes réfléchies à des degrés différents.

À midi, le soleil pénètre à la verticale dans notre atmosphère, et son trajet qu'au sol est relativement court. Le matin et le soir,

has to travel far further. As it enters the atmosphere at an angle, it passes through a larger area of the atmosphere until it reaches our eyes, hitting more particles and being dispersed to a far greater extent.

If there is a particularly large amount of humidity or dust in the atmosphere, the warmer tones of sunlight are reflected more, while the blue wavelengths being filtered out: in these conditions, sunrises and sunsets can appear especially spectacular. Cynics might say that air pollution doesn't only make it more difficult to breathe, but also produces breathtaking displays on the horizon; volcanic eruptions produce even more spectacular sun-ups and sun-downs, however, because they emit huge amounts of particles into the atmosphere which refract the red, orange, and yellow wavelengths in sunlight. Moreover, since volcanic ash is spewed out to height of 100,000 feet, this effect is not limited to the immediate vicinity of the eruption. Depending on the winds, the dust and debris from a volcano can cover thousands of kilometres.

Atmosphäre ein und hat daher bis zum Erdboden einen relativ kurzen Weg. Morgens und abends ist dieser Weg viel länger. Das Licht kommt von der Seite und durchquert einen viel größeren Bereich unserer Atmosphäre, bevor es unser Auge erreicht. Das bedeutet auch, dass das Licht auf seinem Weg auf sehr viel mehr Teilchen trifft und sehr viel stärker daran abgelenkt wird.

Wenn besonders viel Feuchtigkeit und Staub in der Luft ist, werden vor allem der Rot- und Gelbanteil des Sonnenlichts reflektiert. Der blaue Anteil dagegen wird herausgefiltert und verschluckt – gerade in solchen Fällen erscheinen die Sonnenauf- und -untergänge besonders spektakulär. Zynischerweise kann so auch die Luftverschmutzung über Städten morgens und abends atemberaubende Farbspiele erschaffen. In den großartigsten Farben aber leuchtet der Himmel nach Vulkanausbrüchen. Hier werden besonders viele Teilchen in die Atmosphäre geschleudert, an denen sich das Licht brechen kann. Und das nicht nur in direkter Nähe des Vulkans. Vulkanasche wird bei einem Ausbruch 20 bis 30 Kilometer in die Höhe geworfen und – je nach Windstärke – Tausende Kilometer weit getragen.

ce trajet est bien plus long. Son rayonnement est latéral et doit traverser une portion bien plus importante de notre atmosphère avant d'atteindre notre œil. Cela signifie aussi que la lumière rencontre bien plus de particules sur son chemin, qui la dévient encore plus fortement.

Quand il y a beaucoup d'eau et de poussière en suspension dans l'air, celles-ci réfléchissent principalement le rouge et le jaune de la lumière solaire. Le bleu est en revanche filtré, absorbé, ce qui donne des levers et couchers de soleil spectaculaires. Comble du cynisme, la pollution atmosphérique au-dessus des villes peut engendrer des jeux de lumière saisissants à l'aube et au crépuscule. Et c'est après des éruptions volcaniques que le ciel se pare des couleurs les plus magnifiques : le rouge et le jaune de la lumière se brisent alors sur les particules éminemment nombreuses rejetées dans le ciel, et ce, pas uniquement à proximité immédiate du volcan. Les cendres sont projetées à 20-30 kilomètres de hauteur et peuvent, selon la direction des vents, être transportées sur des milliers de kilomètres.

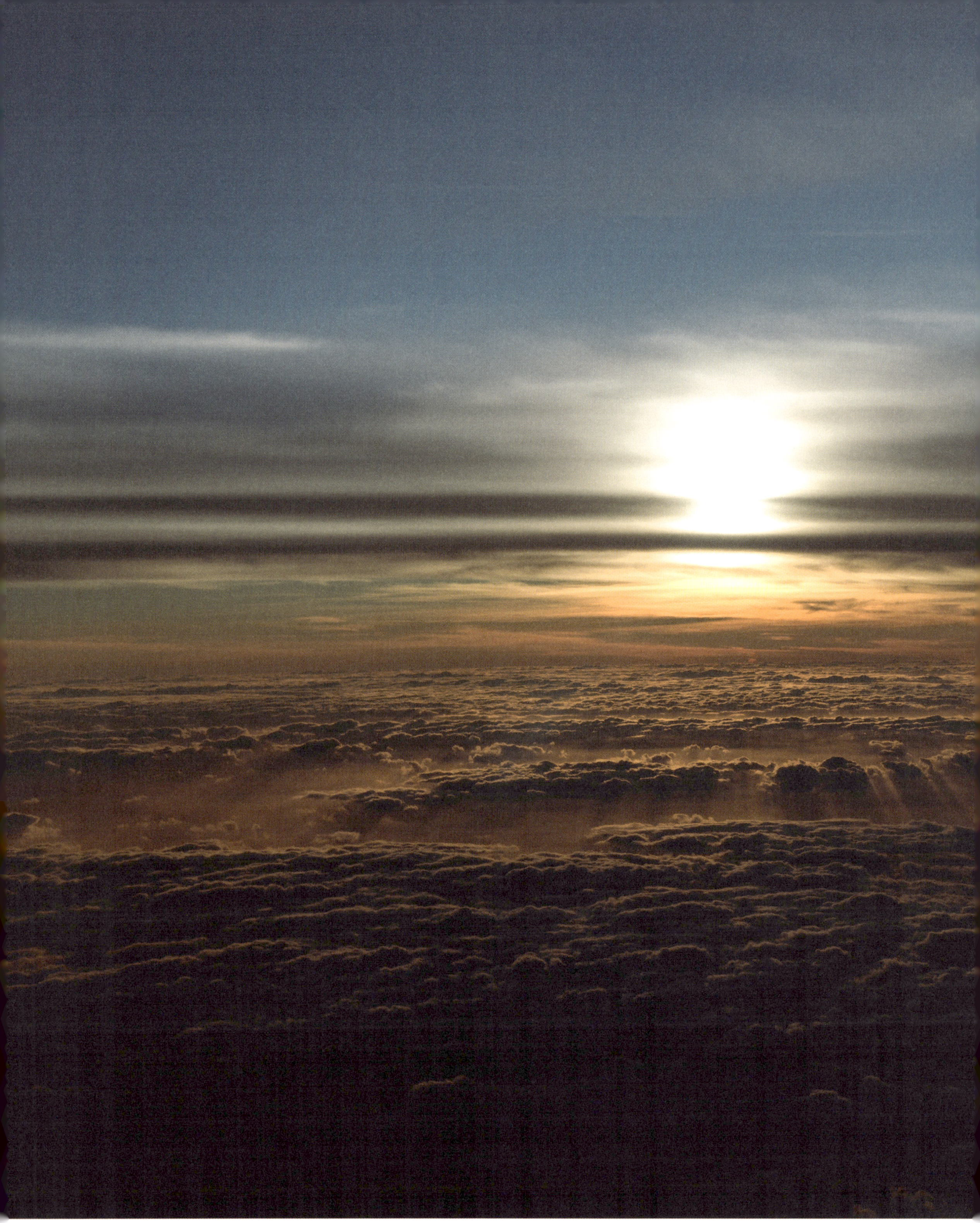

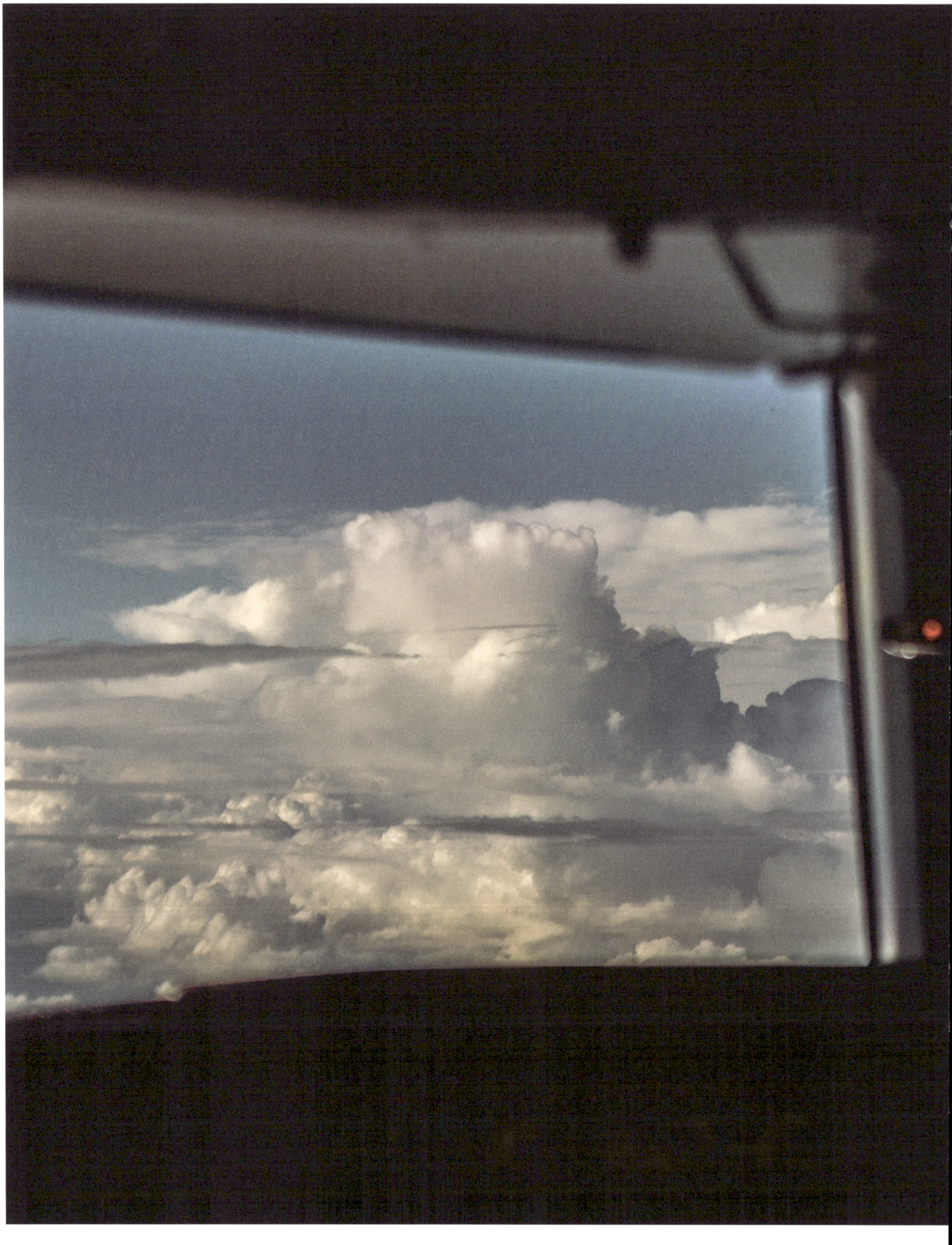

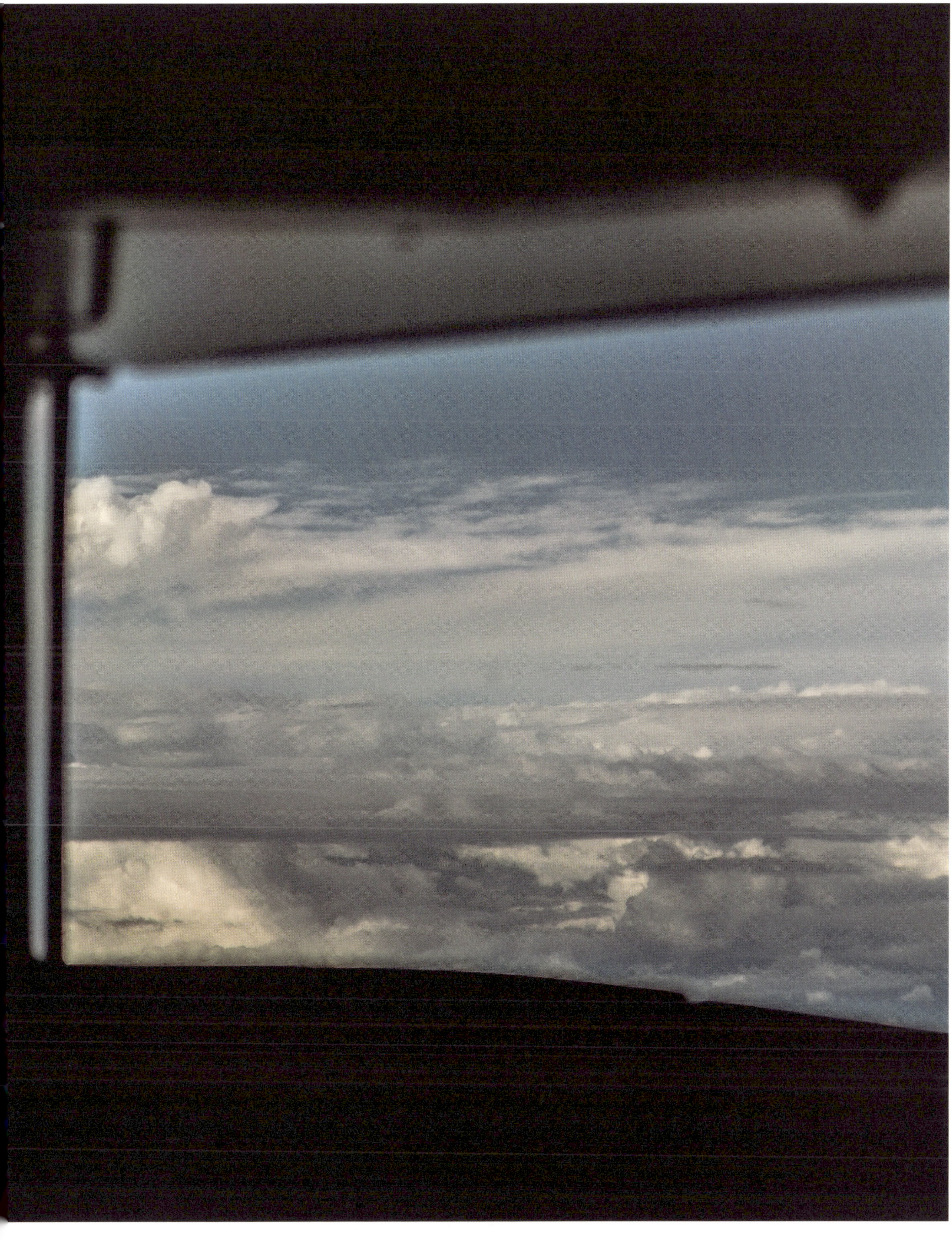

What is a halo?

The word halo comes from the Greek halos and means ring of light; typically, it is used in a religious context, but this optical description is very close to the meteorological definition too, although this latter encompasses both milky rings and other areas of light which appear around the sun or moon in specific conditions. Parhelia, for instance, also known as "sun dogs" or "mock suns", are included in the definition too; these small rainbow-coloured points of light, which appear left and right of the sun or the moon, are just one of the more than 50 light effects grouped together under the name halo. These optical phenomena are the result of ice crystals in the sky, either grouped as thin cirrostratus clouds in the upper atmosphere or in freezing fog closer to the earth's surface. Halos may appear when sunlight is reflected or refracted by these crystals, which means that they are in essence the 'icy versions' of rainbows inasmuch as light is being broken by ice crystals rather than water droplets. For the effect to appear however, the crystals need to be especially regular in form and utterly translucent.

Halos in winter fog when the sun is only just perceptible from within the patch of cloud are particularly impressive. In Europe, there are on average around 100 days a year on which halo phenomena are visible, so it's always worth having a look up towards the sun, especially when high-altitude cirrus clouds are visible.

Was ist ein Halo?

Das Wort „Halo" stammt aus dem Griechischen und bedeutet „Lichthof" – was dem meteorologischen Halo schon sehr nahe kommt. Halos sind Lichteffekte in Form von hellen Höfen oder milchigen Ringen, die unter bestimmten Bedingungen meist um Sonne oder Mond herum zu sehen sind. Auch sogenannte Nebensonnen werden als Halos bezeichnet: kleine, in den Regenbogenfarben leuchtende Punkte rechts und links der Sonne. In der Meteorologie ist Halo der Sammelbegriff für eine ganze Reihe an solchen und ähnlichen Lichteffekten – über 50 verschiedene gibt es davon.

Notwendig für die Entstehung sind vor allem Eiskristalle. Diese existieren in Form von Schleierwolken weit oben am Himmel oder in bodennahem Eisnebel. Halos können sich bilden, wenn sich das Sonnenlicht in den Kristallen bricht oder spiegelt. Sie sind damit das „eisige" Pendant zu Regenbögen, bei denen Licht – statt in Eiskristallen – in Wassertröpfchen abgelenkt wird. Die Kristalle müssen besonders regelmäßig gewachsen und durchsichtig sein, um einen Halo entstehen zu lassen.

Besonders beeindruckend wirken sie im Winter bei Nebel. Wenn die Sonne hinter der Hochnebelsuppe nur noch zu erahnen ist. In Mitteleuropa treten Halo-Erscheinungen an etwa 100 Tagen im Jahr auf – es lohnt sich also, immer mal wieder gen Himmel zu schauen, vor allem wenn Cirruswolken (besonders hohe Wolkenfelder) zu sehen sind.

Qu'est-ce qu'un halo ?

Le mot grec « halo » désigne une aire à battre le grain et par extension toute surface ronde unie, ce qui nous rapproche de la forme de ce phénomène météorologique. Les halos sont des effets de lumière qui ont la forme de foyers lumineux ou d'anneaux blanc laiteux et sont généralement visibles autour du soleil ou de la lune dans des conditions particulières. On appelle aussi halos les parhélies, petits points brillant dans les couleurs de l'arc-en-ciel à gauche et à droite du soleil. En météorologie, halo est un terme générique servant à désigner une bonne cinquantaine d'effets lumineux de ce type ou semblables.

Leur formation requiert avant tout la présence de cristaux de glace. Ces derniers se rencontrent dans les voiles nuageux d'altitude ou dans le brouillard givrant au sol. Les halos se forment quand la lumière solaire se réfracte ou se réfléchit sur les cristaux. Le halo est le pendant « glacé » de l'arc-en-ciel, dans lequel la lumière est déviée, non dans des cristaux de glace, mais dans des gouttelettes d'eau. Les cristaux doivent avoir une forme bien régulière et être transparents pour pouvoir engendrer un halo. Ils sont vraiment impressionnants en hiver par temps de brouillard, quand on ne peut plus que deviner le soleil à l'arrière-plan. En Europe centrale, ces phénomènes étant visibles environ 100 jours par an, cela vaut la peine d'observer le ciel de temps à autre, surtout quand des cirrus le traversent en altitude.

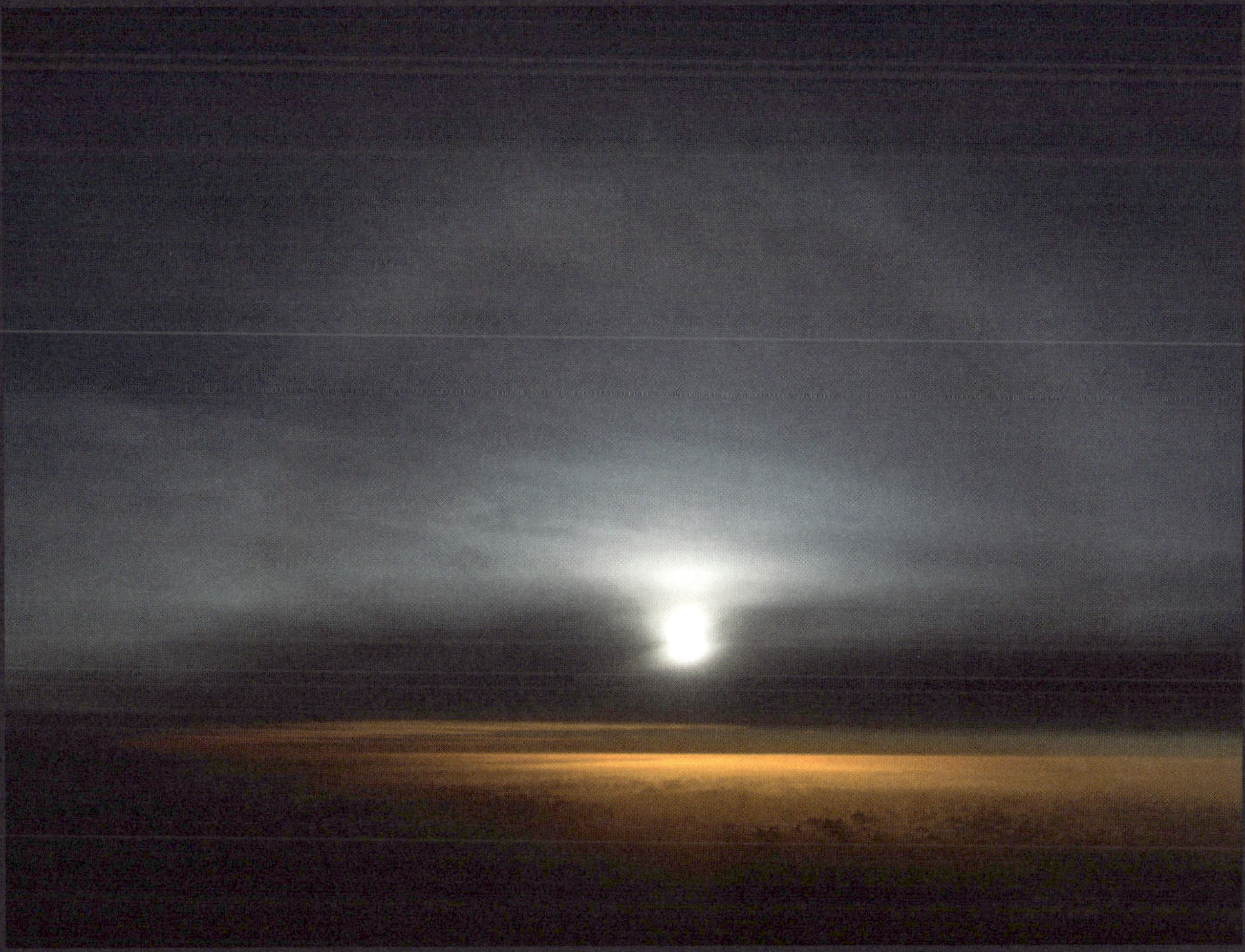

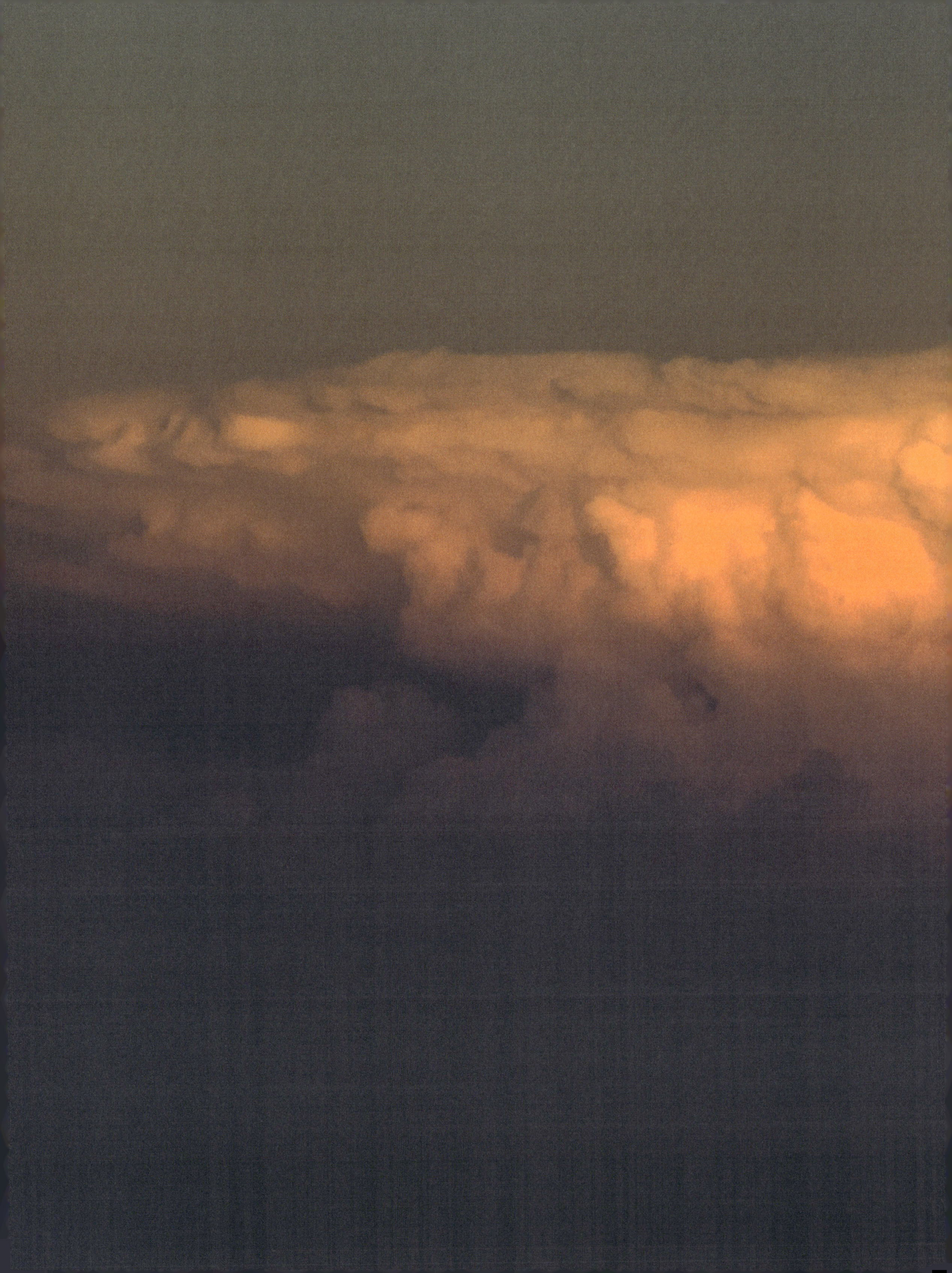

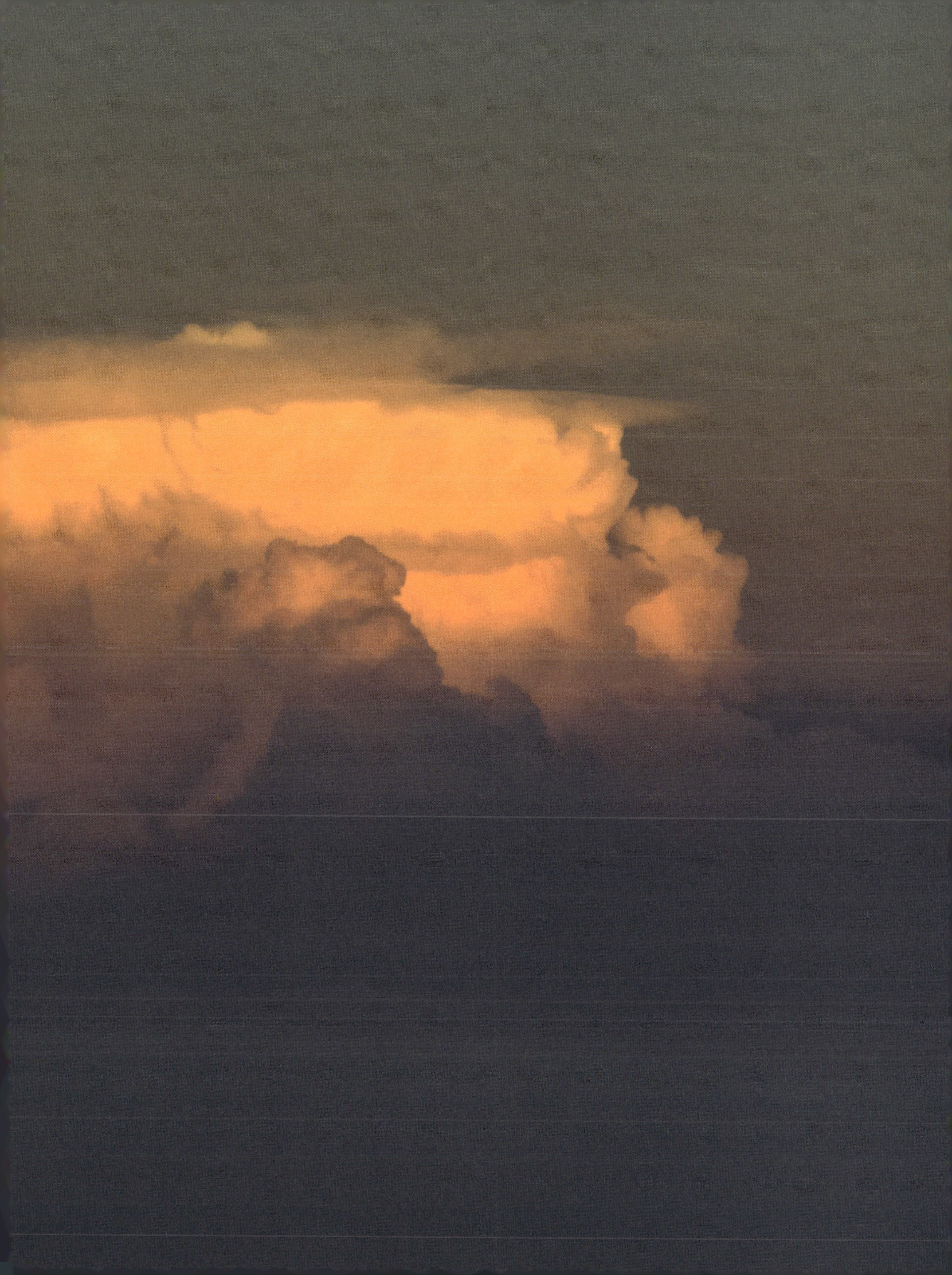

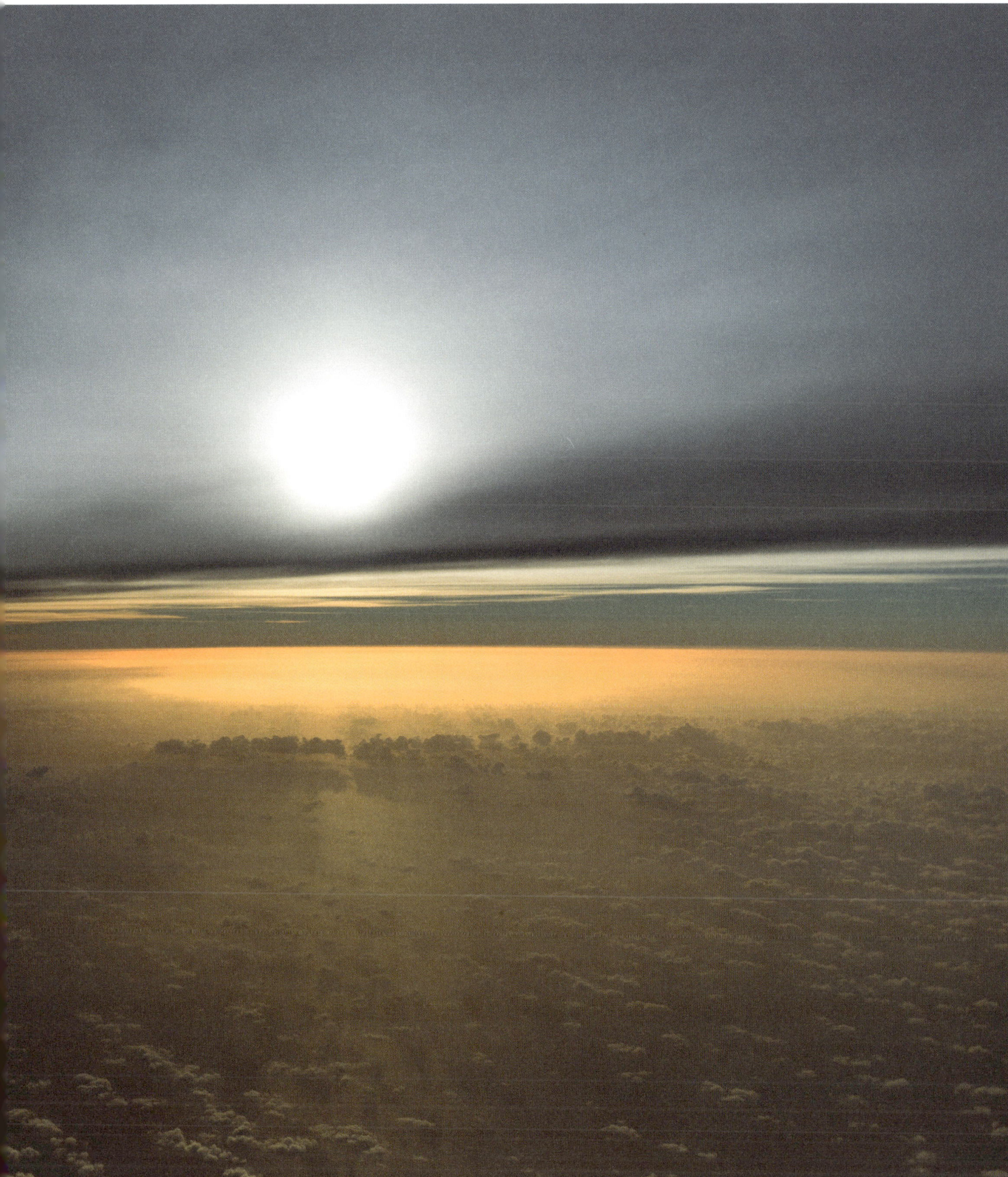

It feels odd how such beautiful landscapes happen #daily right in the middle of the ocean with no one around to admire them

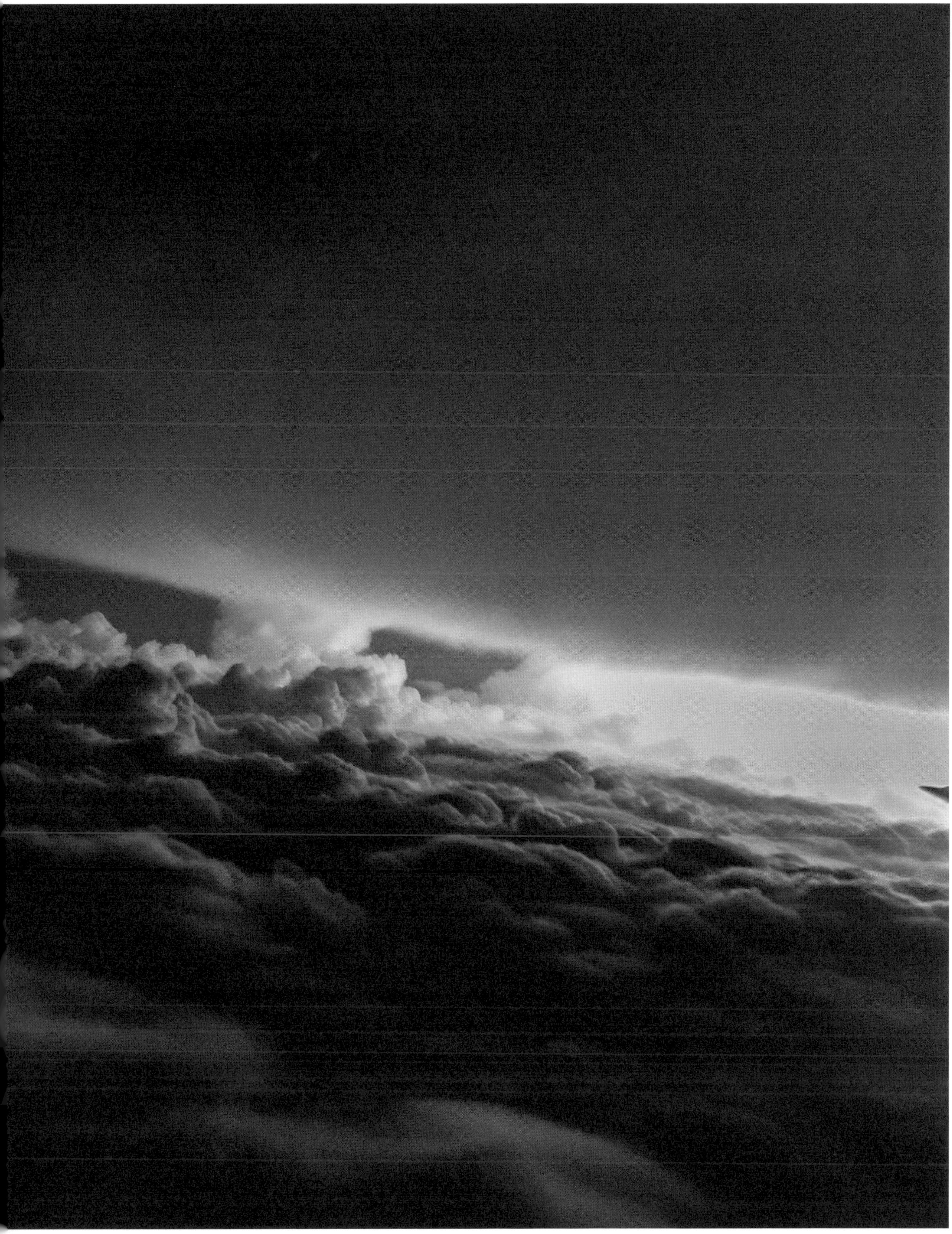

Where does space begin?

The question might be simple, but the answer is complicated—and depends on who you ask. The US Air Force defines anyone who has been 50 miles above sea-level as an astronaut, but this is actually quite an arbitrary threshold which is not supported by a relevant physical principle. NASA, meanwhile, sets the barrier for going cosmic at 100 kilometres (328,000 feet) and there is a recognisable logic behind this: the higher an aircraft climbs, the thinner the atmosphere becomes and the faster it has to fly in order to avoid losing altitude. At around 100 kilometres up however, above what is known as the Kármán Line, the density of the atmosphere is no longer an element in aerodynamic lift: the centrifugal force of the earth is enough to keep the craft in stable orbit.

Nevertheless, this line too, is just one possible boundary among many: there is no one single point at which the atmosphere stops and space begins. Rather, it's more of a transition between the two; even at 250 miles above ground level—the height at which the International Space Station orbits the earth—there are still traces of the atmosphere. Outer space beyond this point is infinite, limitless and (measured against this extent) utterly empty. As well

Wo beginnt der Weltraum?

Simple Frage – schwierige Antwort: Es kommt darauf an, wen man fragt. Die U.S. Air Force definiert jeden als „Astronaut", der sich in einer Höhe von 50 Meilen, umgerechnet also etwa 80 Kilometer über dem Meeresspiegel befindet. Diese Grenze ist allerdings etwas willkürlich, einen guten physikalischen Grund dafür gibt es nicht. Die NASA dagegen legt die Grenze auf 100 Kilometer fest – was schon etwas logischer erscheint. Denn je höher ein Flugzeug steigt, desto dünner wird die Atmosphäre, und desto schneller muss es fliegen, um nicht abzustürzen. Ab circa 100 Kilometern Höhe aber, der sogenannten Kármán-Linie, spielt die Dichte der Atmosphäre keine Rolle mehr und allein die Fliehkraft der Erde reicht aus, um das Flugzeug stabil zu halten.

Dass auch diese Grenze ziemlich willkürlich ist, versteht sich von selbst. Es gibt eben keine einfache Linie, wo sich die Erdatmosphäre vom Weltraum trennt. Vielmehr ist es ein langsamer Übergang. Noch 400 Kilometer über dem Erdboden, dort, wo die ISS ihre Bahnen zieht, lassen sich noch Reste der Atmosphäre finden. Der Weltraum dahinter ist wahrscheinlich unendlich groß, hat keine Grenzen und ist – nimmt man seine Größe als

Où l'espace commence-t-il ?

À cette question simple, il n'y a pas de réponse simple : cela dépend de la personne que l'on interroge. Pour l'armée de l'air américaine, est un astronaute toute personne se trouvant à une altitude de 50 miles, c'est-à-dire environ 80 kilomètres au-dessus du niveau de la mer. Cette limite, que rien ne justifie d'un point de vue physique, est plutôt arbitraire. La NASA, quant à elle, fixe cette limite à 100 kilomètres, ce qui semble déjà plus logique. En effet, plus un avion prend de l'altitude, plus l'atmosphère se raréfie, et plus il doit voler vite pour ne pas décrocher. Or à partir d'une altitude d'environ 100 kilomètres, à savoir de la ligne dite de Kármán, la densité de l'atmosphère ne joue plus aucun rôle et la force centrifuge de la Terre suffit à maintenir l'avion dans les airs.

Il va de soi que cette limite est aussi relativement arbitraire. Le fait est qu'il n'y a pas de ligne de démarcation claire et nette entre l'atmosphère terrestre et l'espace, mais une transition lente. À 400 kilomètres d'altitude, là où l'ISS est en orbite, on trouve encore des résidus d'atmosphère. L'espace qui se trouve au-delà est vraisemblablement infiniment grand, illimité et, comparé à sa taille, presque entièrement vide. En dehors des étoiles, des planètes et autres débris de roche,

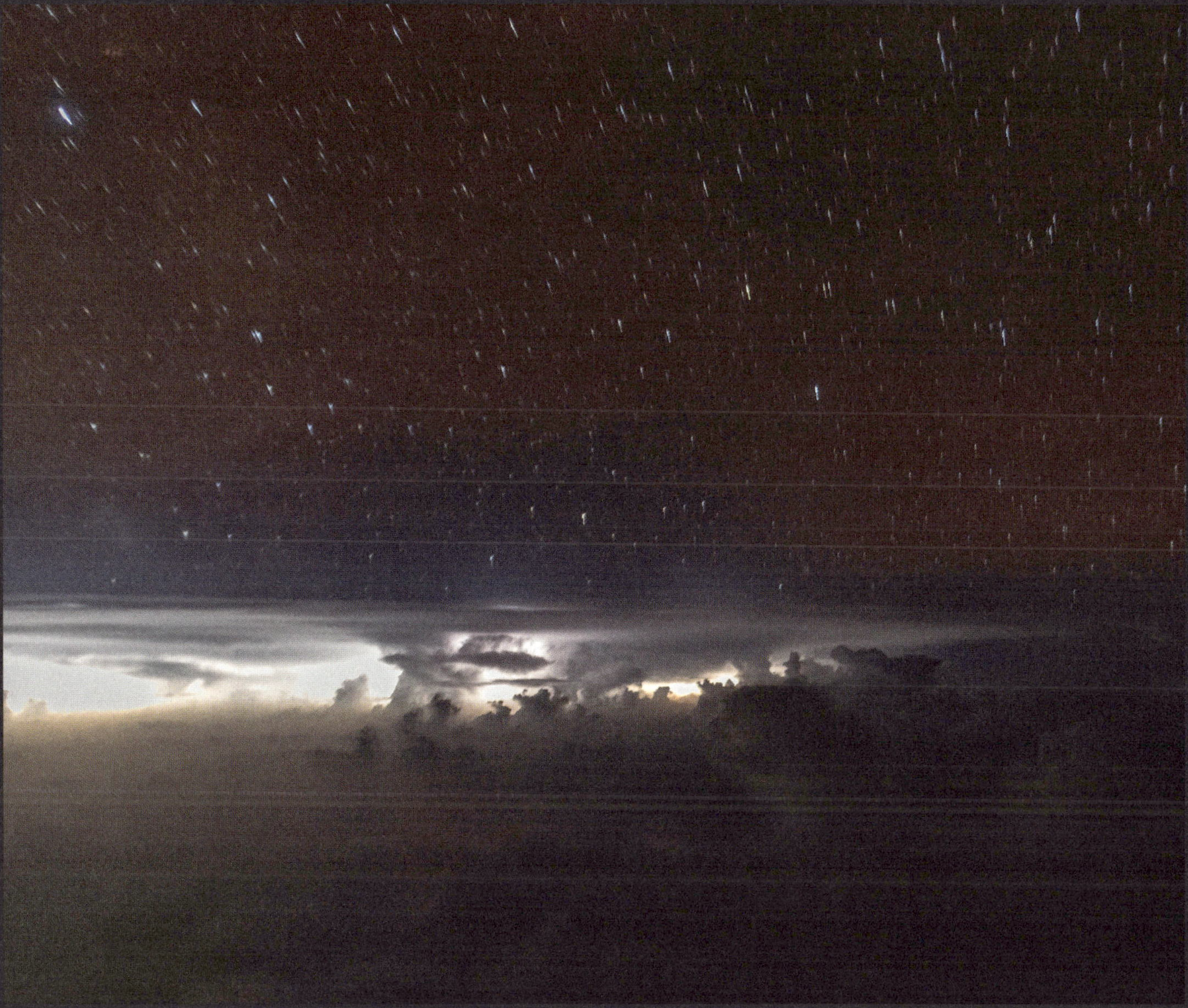

as stars, planets, asteroids and other rocks and boulders, deep space contains little more than tiny particles of dust and very thinly dispersed gases. Space is so expansive that much of the light within it has simply yet to reach us. That is why, on photos, it appears black as pitch.

Maßstab – fast vollkommen leer. Neben Sternen, Planeten und anderen Gesteinsbrocken gibt es hier nur winzige Staubpartikel und sehr dünn verteiltes Gas. Somit kann sich hier der Schall nicht ausbreiten und es ist mucksmäuschenstill. Das Weltall ist so groß, dass uns ein Großteil des Lichts, das sich in ihm befindet, einfach noch nicht erreicht hat. Auf Fotos sieht der Weltraum deshalb rabenschwarz aus.

on n'y trouve que de minuscules particules de poussière et du gaz très diffus. Le son ne pouvant donc pas s'y propager, on n'y entend pas le moindre bruit. L'espace est si grand qu'une grande partie de la lumière qu'il renferme ne nous a pas encore atteints. C'est pourquoi sur les photos l'espace semble noir comme poix.

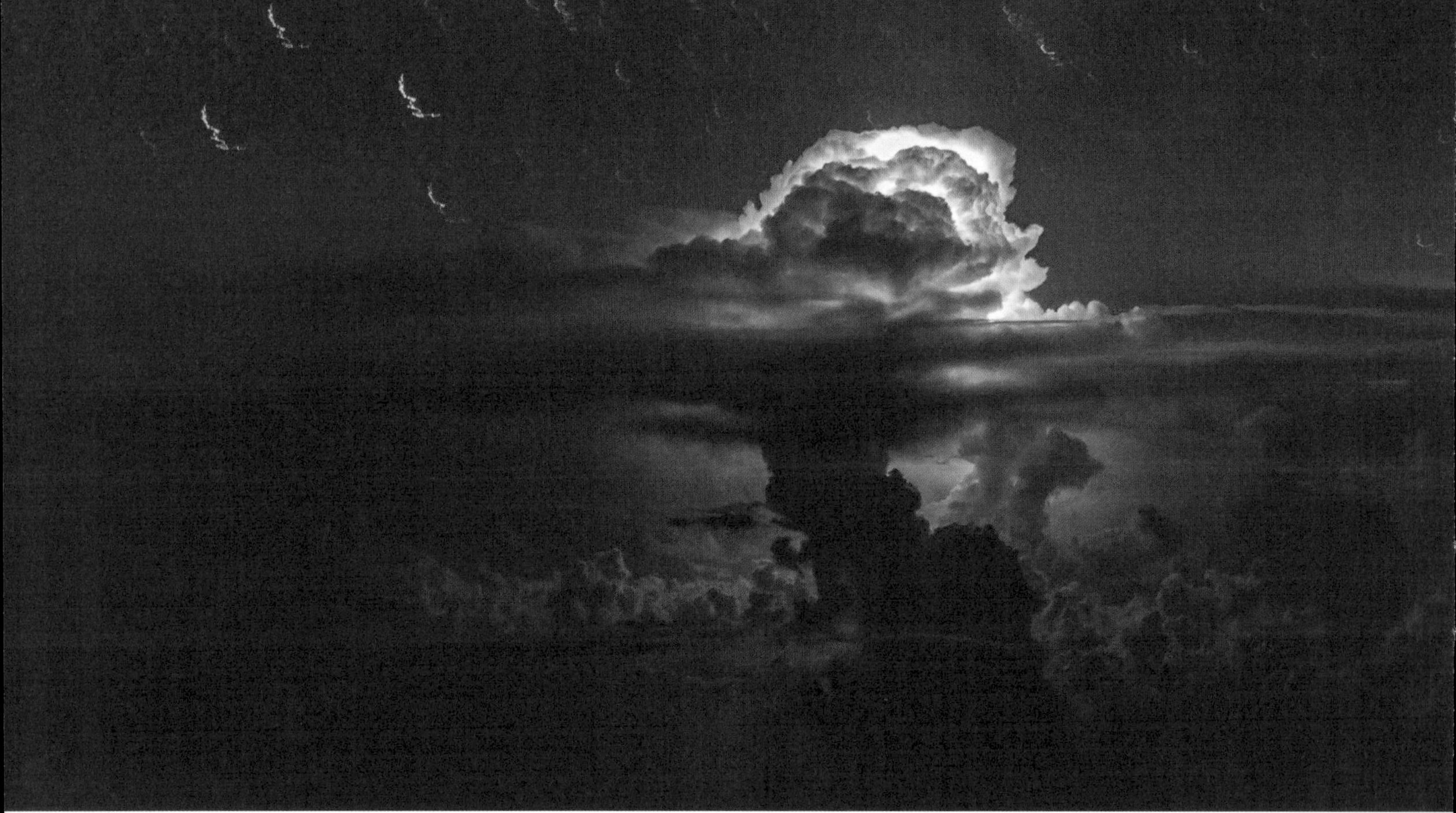

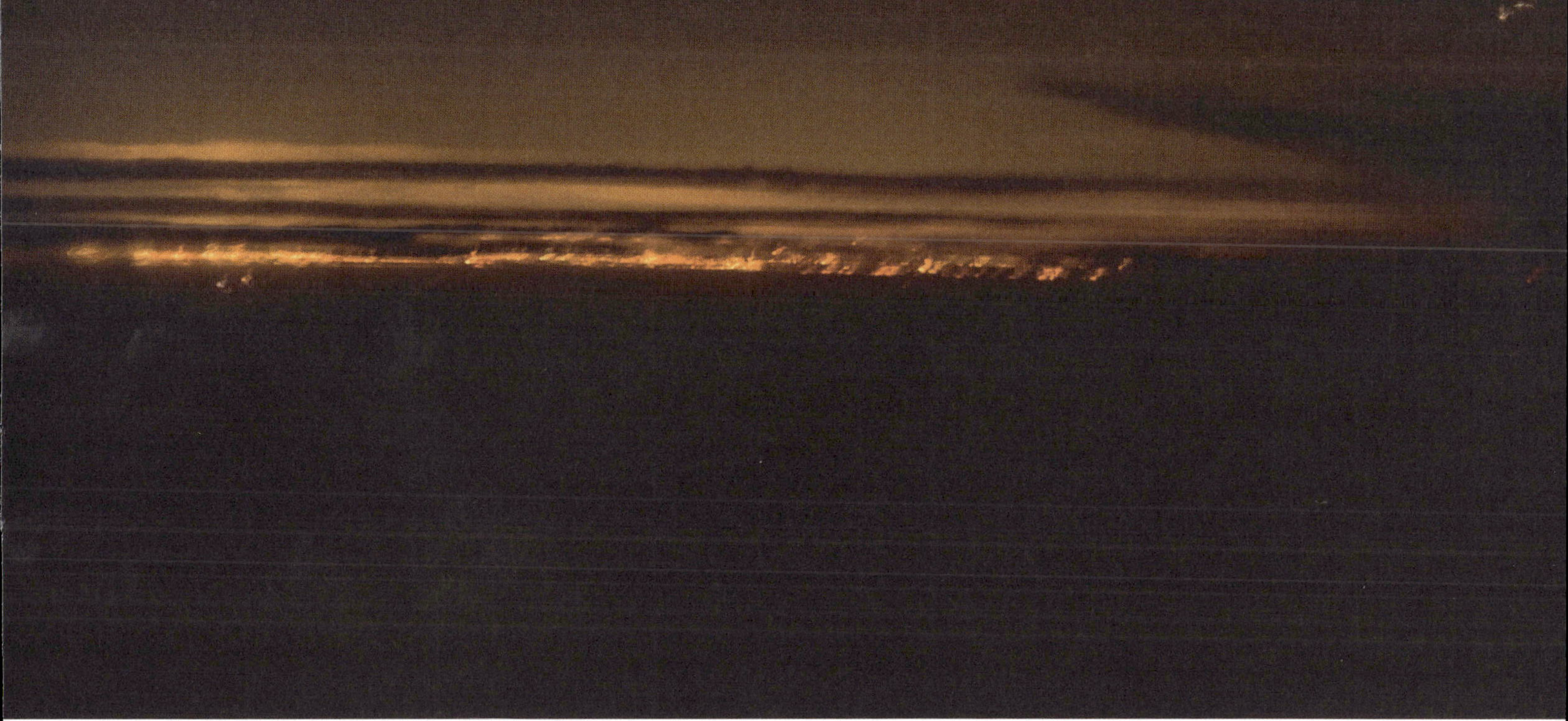

Wondering how do the storms look,
up in the planets around those stars

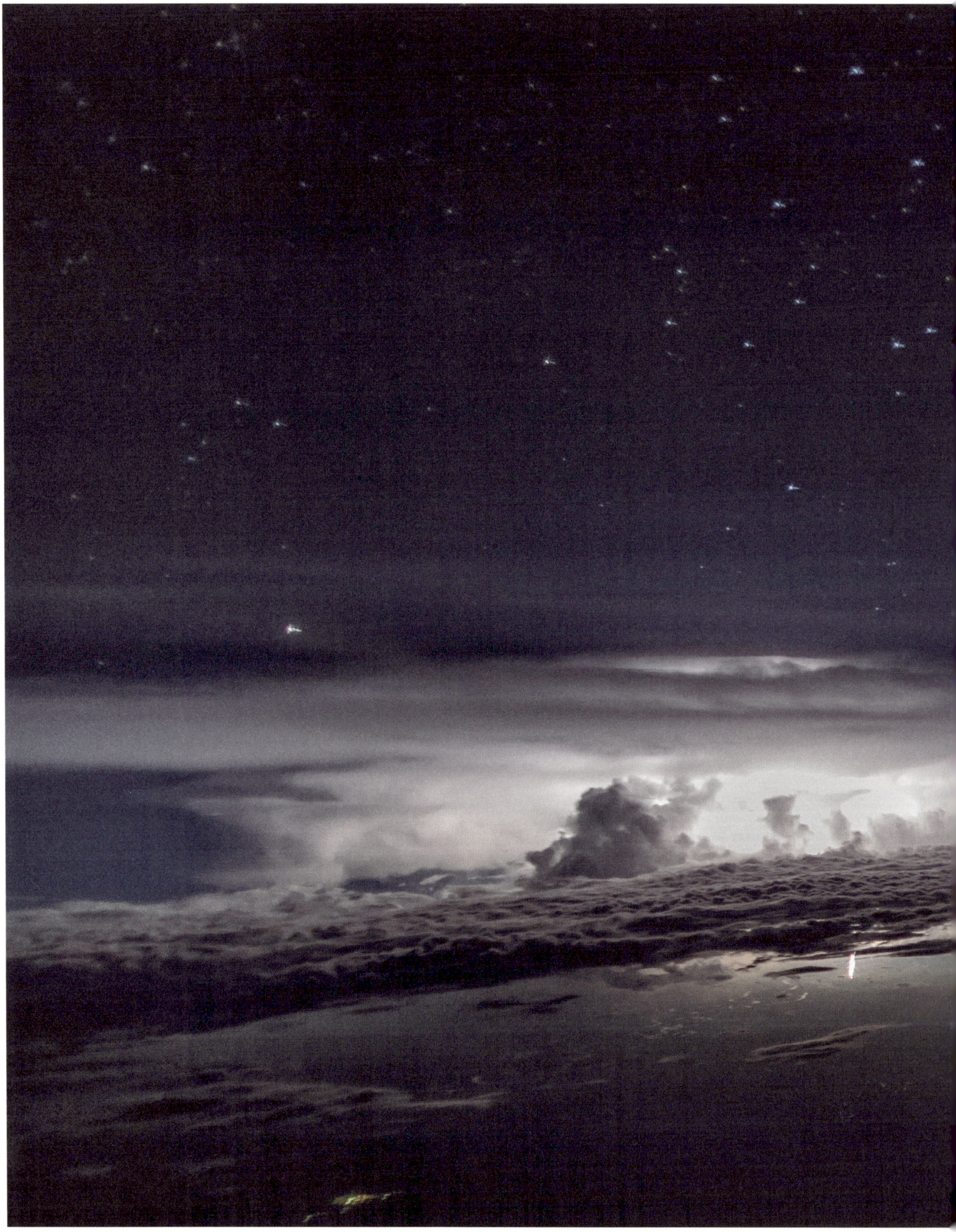

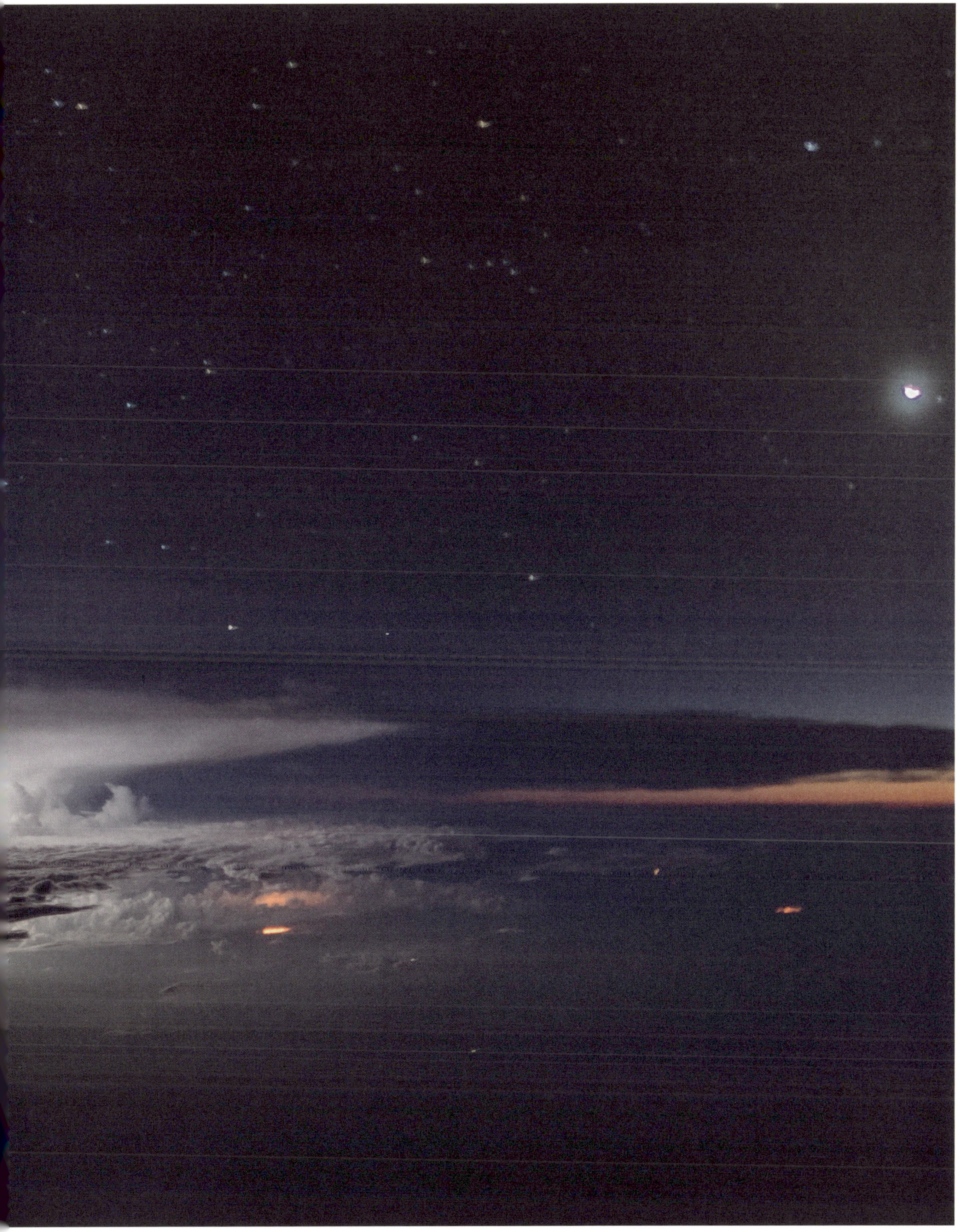

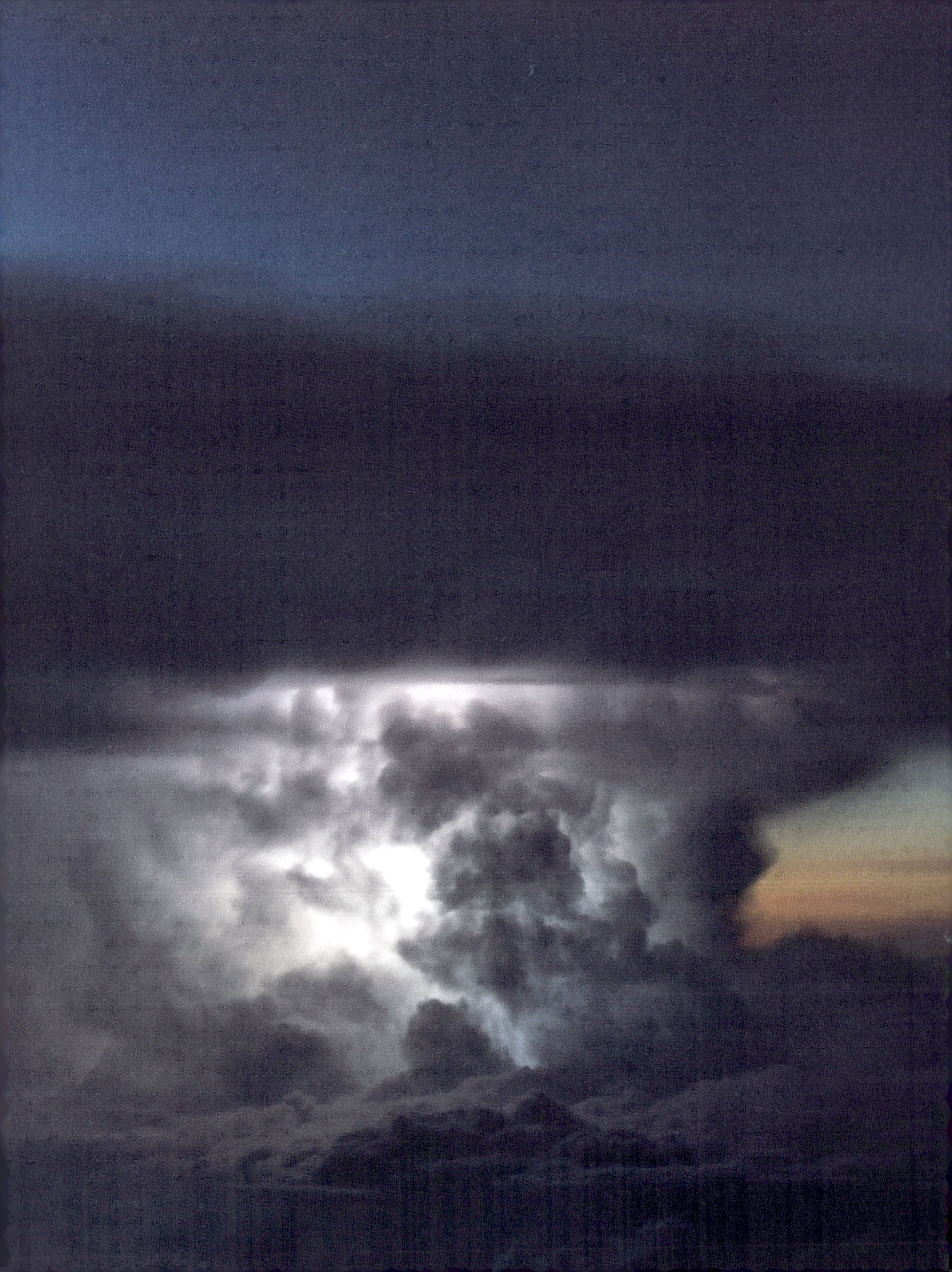

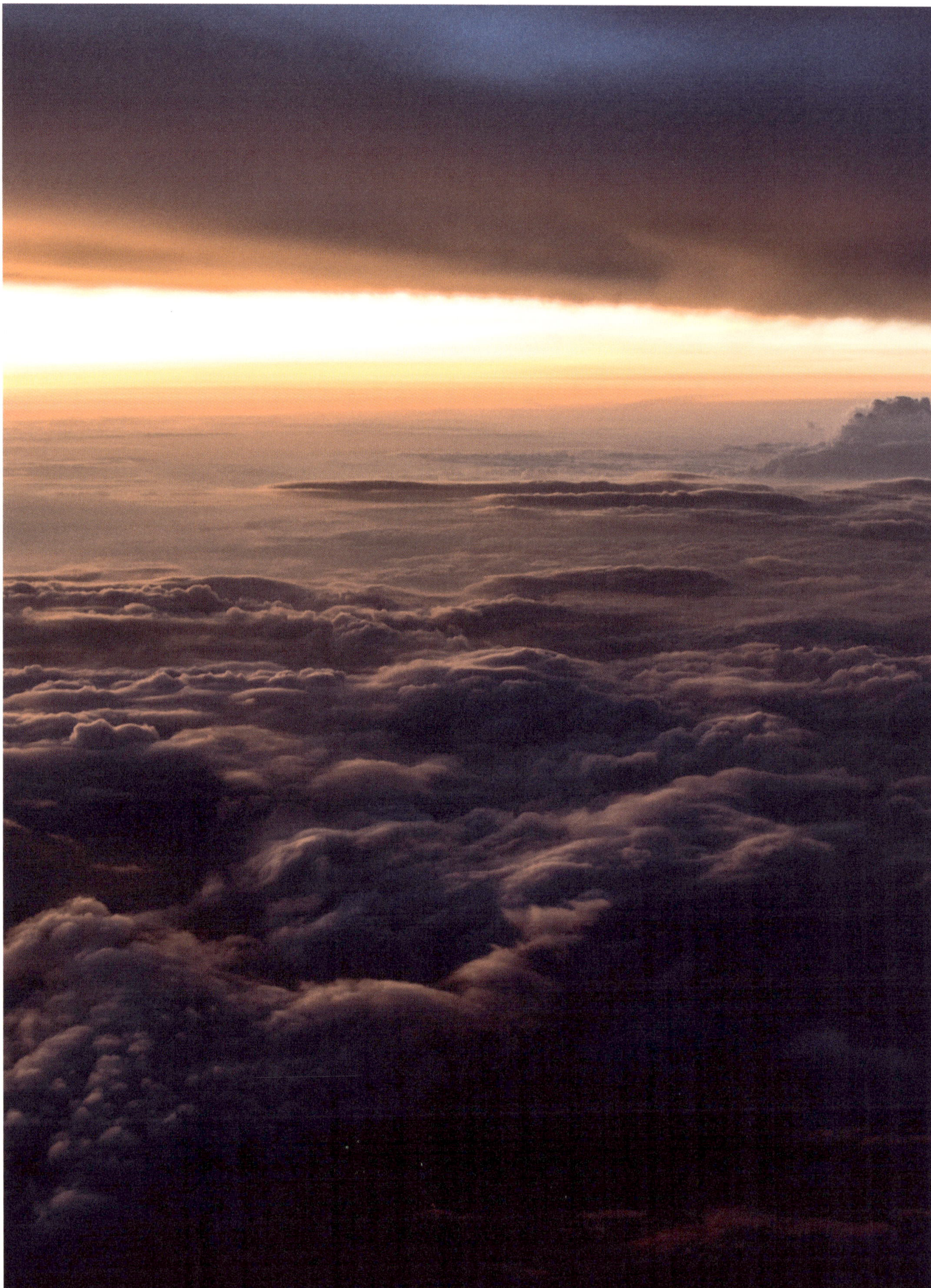

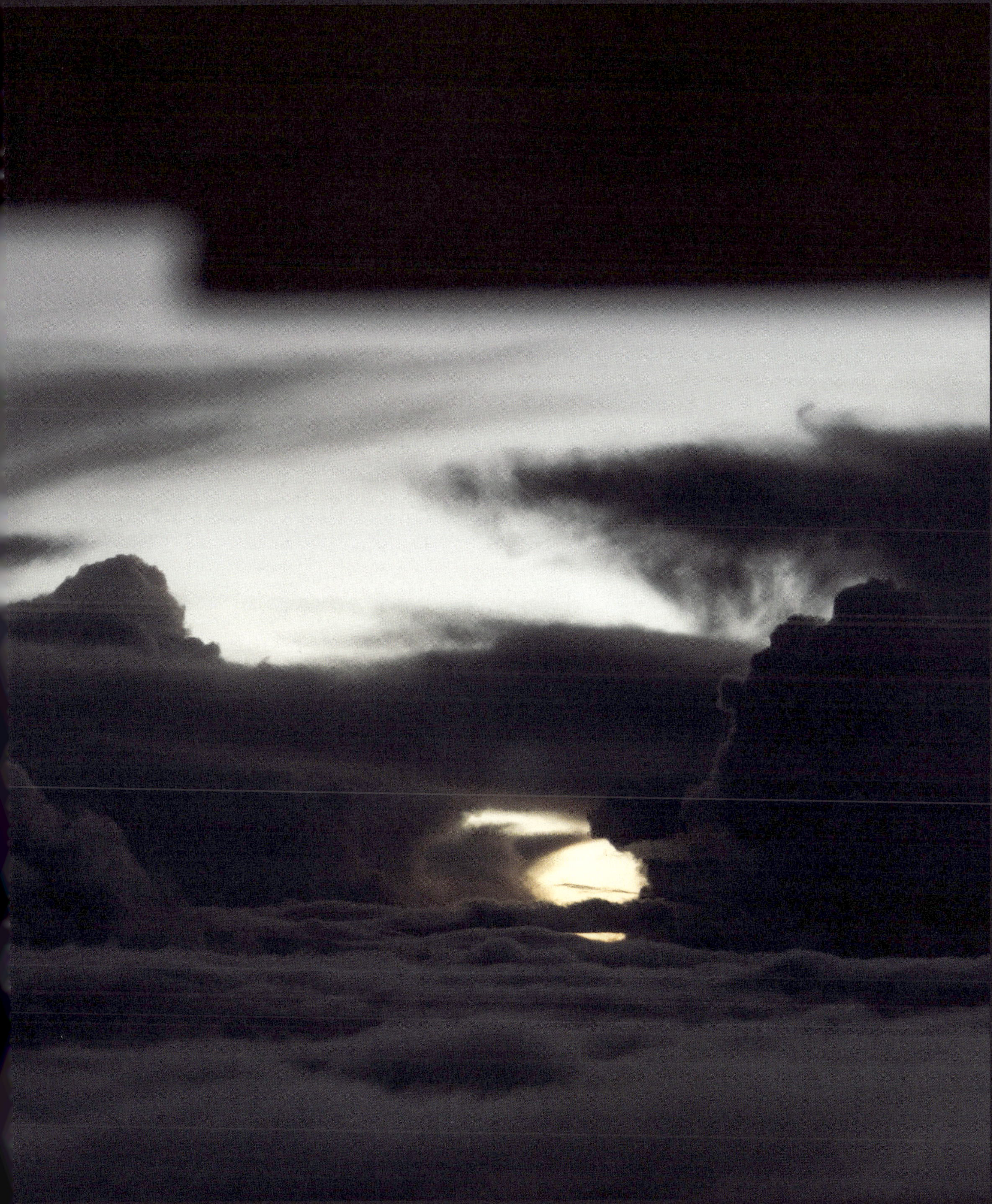

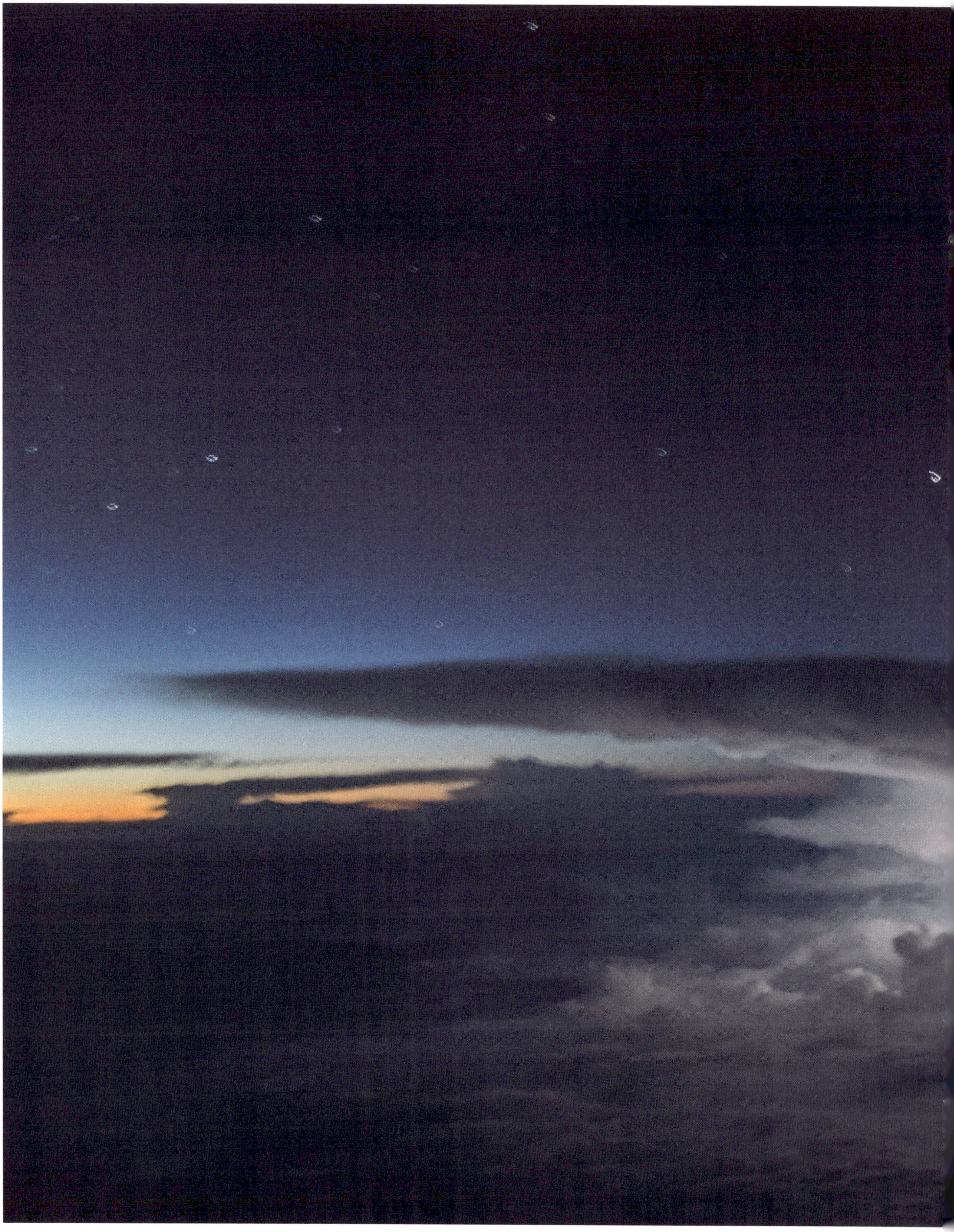

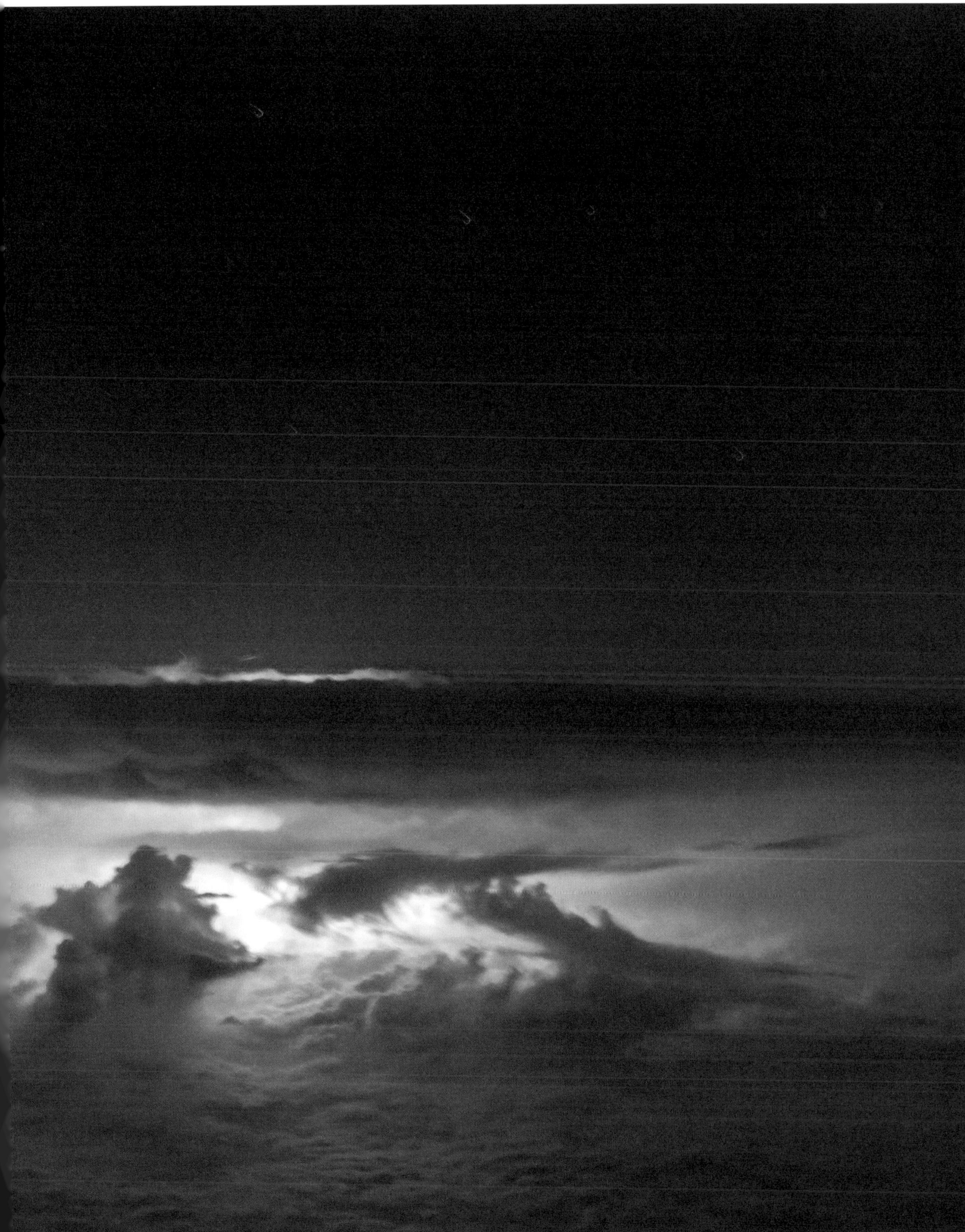

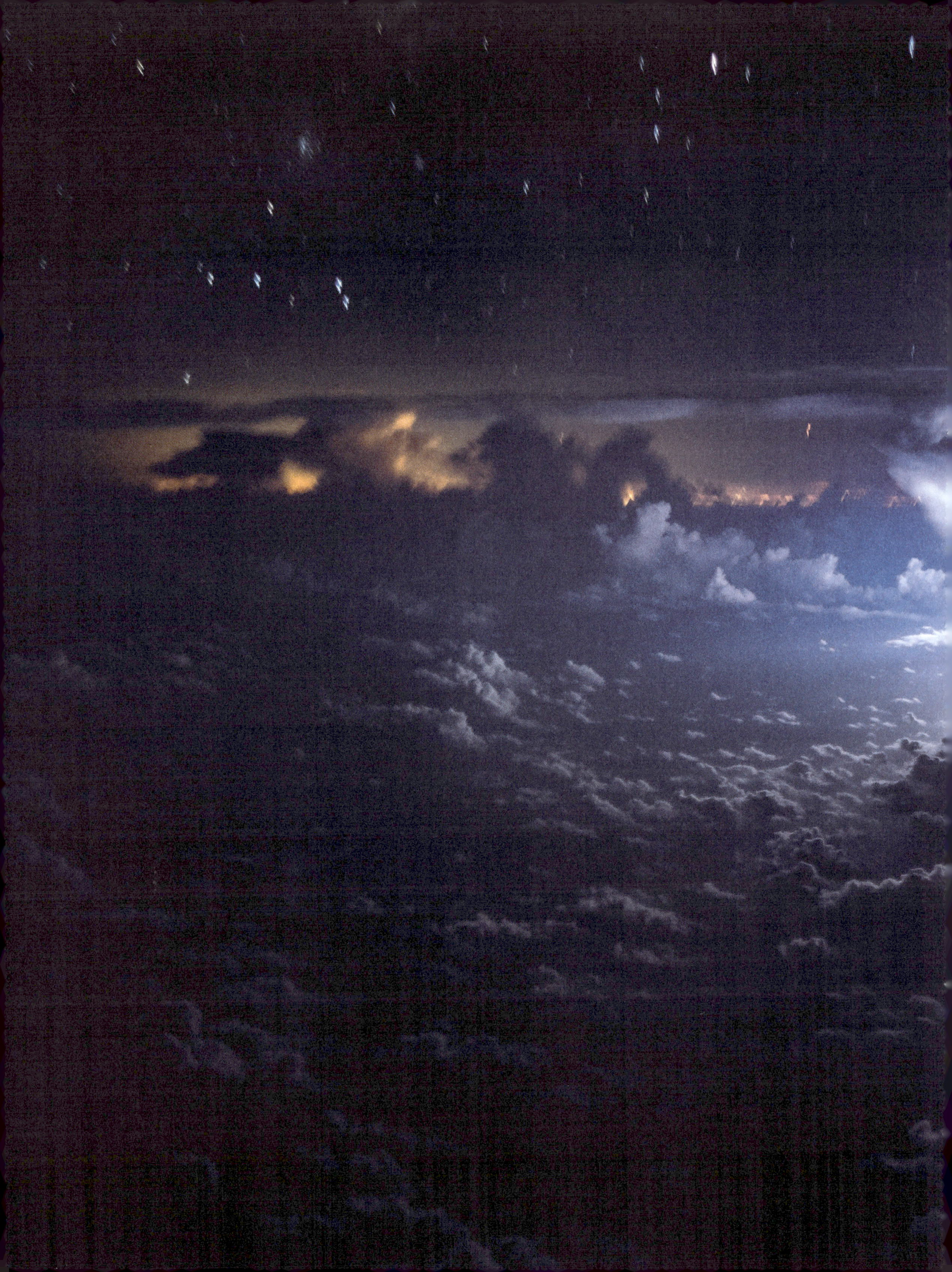

Curriculum vitae

Santiago Borja, born in Quito, Ecuador, works as a pilot for an international airline. Since 2012, he has been first officer on a Boeing 767-300ER. The photographs he takes from the cockpit have won him a range of international prizes and are to be found in galleries worldwide. He lives in Quito.

Biografie

Santiago Borja, geboren in Quito, Ecuador, ist Pilot einer internationalen Airline. Seit 2012 fliegt er als erster Offizier eine Boeing 767-300ER. Seine Fotografien aus dem Cockpit haben zahlreiche internationale Preise gewonnen und hängen in großen Galerien weltweit. Er lebt in Quito.

Vita

Née à Quito, en Équateur, **Santiago Borja** travaille comme pilote pour une comapgnie aérienne internationale. Depuis 2012, il est pilote en premier d'un Boeing 767-300ER. Distinguées par de nombreuses récompenses internationales, les photographies qu'il prend de son cockpit sont exposées dans de grandes galeries du monde entier. Santiago Borja vit à Quito.

Michaela Koschak was born in Berlin in 1977. She studied meteorology at the Free University of Berlin. In 1998, she co-founded a company through which, until 2000, she produced the weather segments for the German private television network SAT.1. She has been anchoring various weather segments on German television since 2003, including the program Kinderwetter (basically, "weather for kids"). Michaela Koschak has authored several children's books about the weather. She lives in Leipzig, Germany.

Michaela Koschak, geboren 1977 in Berlin, studierte Meteorologie an der FU Berlin. 1998 war sie Mitbegründerin einer Wetterfirma, mit der sie bis 2000 die Wetterberichte für den deutschen Privatsender SAT.1 produzierte. Seit 2003 moderiert sie verschiedene Wetterberichte im deutschen Fernsehen, darunter das Format „Kinderwetter". Michaela Koschak hat mehrere Kinderbücher zum Thema Wetter geschrieben und lebt in Leipzig, Deutschland.

Née en 1977 à Berlin, **Michaela Koschak** a étudié la météorologie à l'université libre de Berlin. En 1998, elle a cofondé une entreprise, dans le cadre de laquelle elle a produit jusqu'en 2000 des bulletins météo pour SAT.1, chaîne de télévision allemande privée. Depuis 2003, elle présente les prévisions météorologiques sur diverses chaînes et dans différents formats, notamment l'émission pour enfants « Kinderwetter ». Michaela Koschak a par ailleurs écrit plusieurs documentaires jeunesse sur le thème de la météo. Elle vit à Leipzig, en Allemagne.

Foreword: Santiago Borja
Text: Michaela Koschak
Translations: Brian Melican (English)
Christèle Jany, Claude Checconi (French)
Creative Director: Martin Graf
Design & Layout: Robin John Berwing
Editorial coordination: Roman Schmid
Production: Nele Jansen
Color separation: Robert Kuhlendahl

ISBN 978-3-96171-108-6

Library of Congress Number: 2018904692

Printed in the Czech Republic

Bibliographic information published by
the Deutsche Nationalbibliothek
The Deutsche Nationalbibliothek lists this publication in the Deutsche Nationalbibliografie; detailed bibliographic data are available on the Internet at http://dnb.dnb.de.

Published by teNeues Publishing Group

MIX
Papier aus verantwortungsvollen Quellen
Paper from responsible sources
FSC® C005833

teNeues Media GmbH & Co. KG
Am Selder 37, 47906 Kempen, Germany
Phone: +49-(0)2152-916-0
Fax: +49-(0)2152-916-111
e-mail: books@teneues.com

Press department: Andrea Rehn
Phone: +49-(0)2152-916-202
e-mail: arehn@teneues.com

Munich Office
Pilotystraße 4, 80538 Munich, Germany
Phone: +49-(0)89-443-8889-62
e-mail: bkellner@teneues.com

Berlin Office
Kohlfurter Straße 41–43, 10999 Berlin, Germany
Phone: +49-(0)30-4195-3526-23
e-mail: ajasper@teneues.com

teNeues Publishing Company
350 7th Avenue, Suite 301, New York, NY 10001, USA
Phone: +1-212-627-9090
Fax: +1-212-627-9511

teNeues Publishing UK Ltd.
12 Ferndene Road, London SE24 0AQ, UK
Phone: +44-(0)20-3542-8997

teNeues France S.A.R.L.
39, rue des Billets, 18250 Henrichemont, France
Phone: +33-(0)2-4826-9348
Fax: +33-(0)1-7072-3482

www.teneues.com

teNeues Publishing Group
Kempen
Berlin
London
Munich
New York
Paris

teNeues